PHYSICAL SCIENCE
Investigating Matter and Energy

PHYSICAL SCIENCE
Investigating Matter and Energy

By

Robert H. Marshall

and

Donald H. Jacobs

Media Materials, Inc.
Baltimore, Maryland 21211

AUTHORS

Robert H. Marshall, M.Ed.
Science Teacher
Baltimore City Public Schools
Baltimore, Maryland

Instructor in Microcomputers
Baltimore County Public Schools
Towson, Maryland

Instructor in BASIC Programming
The Stratford Schools
Baltimore, Maryland

Co-author of:
 Basic Steps to BASIC

Donald H. Jacobs, M.Ed.
Mathematics Teacher
Baltimore City Public Schools
Baltimore, Maryland

Microcomputer Lab Coordinator
Talmudical Academy
Pikesville, Maryland

Mathematics Coordinator
Educational Resource Services
Pikesville, Maryland

Co-author of:
 Basic Mathematics Skills
 Life Skills Mathematics
 Basic Steps to BASIC

CONSULTANTS

William J. Brennan, B.S.
Science Teacher
Baltimore City Public Schools

Carolyn R. Newsome, M.S.
Science Teacher
Baltimore City Public Schools

Willette Harbor, B.S.
Science Teacher
Baltimore City Public Schools

Editor: Barbara Pokrinchak, Ed.D.
Editorial Consultant: M. E. Criste
Art: Gregory Broadnax

ISBN: 0-86601-563-9   05 87 VG/ATB   5.0   Printed in the U.S.A.

Contents

List of Investigations

A Word about Computer Programs:
Each of the chapters in this text has a computer program at the end of the chapter. The programs have been written so that they will work properly on most of the microcomputers that are available today. The programs are all written in standard BASIC, the Beginner's All-purpose Symbolic Instruction Code.

Before typing the programs on your computer, read the instruction manual that comes with the computer.

In addition, please read the following notes about all of the programs in the text.

1. The command CLS is used in the programs to clear the screen. This command on your computer may be different. Many computers use the command HOME to clear the screen, while others may use some other word. Use the one which is used in your computer's manual.

2. The PRINT command can be replaced with the question mark (?). This feature is available on most computers and will save you time when typing the programs. To do this, instead of typing:

    10 PRINT "WHAT IS THE VOLUME?"

    you may simply type:

    10 ? "WHAT IS THE VOLUME?"

3. Some computers do not allow the use of multiple statements on a line. For example, the following is not permitted on some computers:

    10 PRINT:PRINT

    The colon (:) is called a statement separator. If your machine does not allow this, you may change the programs as in the following example:

    Change:
        10 PRINT "WHAT IS THE VOLUME?":PRINT
        20 INPUT V

    To: 10 PRINT "WHAT IS THE VOLUME?"
        15 PRINT
        20 INPUT V

    The PRINT statement after the colon in line 10 has been placed on a separate numbered line by itself.

Preface

Physical Science: Investigating Matter and Energy is a primary text for an introductory course covering the physical sciences. It is designed for students who need a simplified presentation of the main concepts of physics and chemistry.

Chapter 1 is an introduction to the use of the metric system of measurement, which is so widely used in scientific studies. The four basic measurements of length, area, volume, and mass are emphasized.

Chapters 2 through 6 cover the properties, structure, classification, and reactions of matter. The examples of matter that are given are common ones with which the students are likely to be familiar.

Chapters 7 through 12 cover the traditional subjects of physics — force, motion, work, heat, sound, light, electricity, and magnetism.

Throughout the book, an effort has been made to use only easily obtainable materials. Many of the activities can be extended by encouraging students to investigate further at home.

SPECIAL FEATURES
- The large type and controlled vocabulary help make the book easy to read.

- Chapter goals are included at the beginning of each chapter so students will know what they will be expected to learn.

- New vocabulary words are written in italics to highlight them for review.

- A glossary of major science terms is included to provide a convenient science vocabulary source.

- Each chapter includes a summary at the end, highlighting the major concepts and terms of the chapter. This provides a quick review of the chapter.

- Review exercises are included for each chapter. This can serve as a formative test for those teachers using the mastery learning method.

- Calculator and/or computer exercises are included for each chapter to provide enrichment or review.

- The *Teacher's Guide* that accompanies this text contains objectives, teaching suggestions, and complete answers. There is a substantial section of resource and practice material.

- The, accompanying student *Manual of Laboratory and Language Activities* provides additional opportunities for skill development in manipulation, observation, measurement, description, and classification.

- An accompanying *Workbook* contains reading and vocabulary exercises that promote fuller comprehension of scientific concepts.

The Metric System

Chapter Goals:

1. To measure the length of objects in meters, centimeters, and millimeters.

2. To convert metric units in meters to centimeters and millimeters.

3. To state the abbreviations for *meter*, *centimeter*, and *millimeter*.

4. To add, subtract, multiply, and divide metric units of measurement.

5. To calculate area and volume.

6. To describe the units of liquid volume and units of mass in the metric system.

The study of science requires the use of many different skills and mathematical operations. One such skill is using the metric system of measurement. The metric system is easy to learn because it is based on multiples of the number 10.

1.1 MEASURING LENGTH

The Meter. The *METER* is used to measure distances in the metric system. A meter is a little more than 39 inches in length, about the length of an average golf club. The meter could be used to measure the length of your yard or the height of your house.

The Centimeter. For some applications, the meter is too large a unit. For example, it would be difficult to measure the wing span of a butterfly using the meter. Instead, the centimeter would be used. A *CENTIMETER* is 1/100 of a meter. It is found by dividing the meter into 100 equal parts. There are 100 centimeters in a meter. The prefix, *centi-*, means 1/100.

The figure below shows how a piece of copper wire is measured with a metric ruler.

Each large numbered space on the ruler represents one centimeter. The wire is 6 spaces long, or 6 centimeters long.

See how the wire has been moved in this figure:

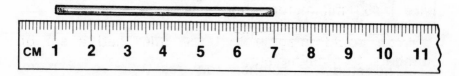

Even though the end of the wire is now near the 7-centimeter mark, the wire is still 6 centimeters in length. Moving the wire does not change its length. The proper way to measure an item is to line up the left edge of the item with the left end of the ruler. Some rulers have a "0" mark on the left edge. In this case, you should begin measuring at the 0 mark.

The Millimeter. A centimeter can be divided into ten smaller units. The ten equal parts that result are called *MILLIMETERS*. The prefix *milli-* in *millimeter* means 1/1,000. There are 1,000 millimeters in one meter.

Study the figure below. There are 10 millimeters for each centimeter.

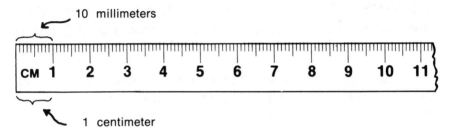

How long is this copper wire in millimeters?

To measure with millimeters, we multiply the number of centimeters by 10. The wire is 6 centimeters long. Multiply 6 centimeters by 10. $6 \times 10 = 60$. The wire is 60 millimeters long.

You could imagine that the ruler had a set of numbers for millimeters:

Study the example below. How long is the wooden match stick in millimeters?

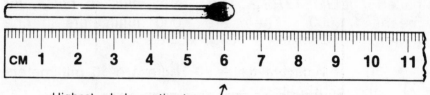

Highest whole centimeter ⟋

Step 1: Read the highest whole centimeter and multiply by 10.

$$6 \times 10 = 60 \quad \text{Product}$$

Step 2: Count the small segments remaining. Add them to the product in step 1.

$$3 + 60 = 63 \text{ millimeters}$$

The match stick is 63 millimeters in length.

The length of the wooden match could also be given in both centimeters and millimeters. You do not have to multiply in step 1. You just read the highest whole centimeter. Then count the remaining millimeters. Express the measurement as 6 centimeters, 3 millimeters. This can be abbreviated to 6 cm 3 mm.

Abbreviations for Metric Units

meter .... m
centimeter .... cm
millimeter ....mm

1.2 CONVERTING METRIC UNITS

+--------------------------------------+
| **Equivalents** |
| |
| 1 meter = 100 centimeters |
| |
| 1 centimeter = 10 millimeters |
+--------------------------------------+

To convert large units to small units, multiply.

Example: 3 cm = ____mm

Solution: 3 × 10 = 30, so 3 cm = 30 mm.

▶ **Practice.** Number your paper from 1 to 16. Write the correct answer next to each number.

1) 3 m = ____cm   2) 5 m = ____cm

3) 7 m = ____cm   4) 2 m = ____cm

5) 6 cm = ____mm   6) 3 cm = ____mm

7) 8 cm = ____mm   8) 5 cm = ____mm

9) 83 millimeters = ____cm ____mm

10) 28 millimeters = ____cm ____mm

▶ Measure each line segment in the units shown.

11) A_____B   ____cm ____mm

12) C_____D   ____mm

13) E_____F   ____cm ____mm

14) G_____H   ____mm

15) I_____J   ____cm ____mm

16) K_____L   ____mm

1.3 WRITING QUANTITIES AS DECIMALS

Often a measurement is a combination of large units and small units. We can write the measurement in the large units by using a decimal.

Example: Write 8 meters 7 centimeters as meters only. Use decimal notation.

Step 1: Write the quantity. Then under that, write the equivalent, so that the numbers form a ratio.

$$\frac{8 \text{ meters}}{1 \text{ meter}} = \frac{7 \text{ centimeters}}{100 \text{ centimeters}}$$

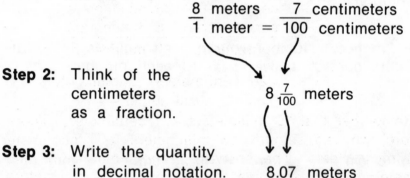

Step 2: Think of the centimeters as a fraction.

$8\frac{7}{100}$ meters

Step 3: Write the quantity in decimal notation.

8.07 meters

▶ *Practice.* Write these measurements as decimals.

1) 6 centimeters 5 millimeters (1 cm = 10 mm)
2) 8 centimeters 3 millimeters (1 cm = 10 mm)
3) 5 meters 8 centimeters (1 m = 100 cm)
4) 6 meters 28 centimeters (1 m = 100 cm)
5) 8 meters 5 millimeters (1 m = 1,000 mm)
6) 19 meters 7 millimeters (1 m = 1,000 mm)

1.4 ADDING AND SUBTRACTING UNITS OF LINEAR MEASURE

You may need to add or subtract linear measurements with different units. Be sure to add like units to like units and to subtract like units from like units.

Example A: Add 6 meters 3 centimeters to 8 meters 4 centimeters.

6 meters	3 centimeters
+ 8 meters	4 centimeters
14 meters	7 centimeters

Example B: Add 9 centimeters 8 millimeters to 10 centimeters 5 millimeters.

9 centimeters	8 millimeters
+10 centimeters	5 millimeters
19 centimeters	13 millimeters, or

20 centimeters 3 millimeters

Remember that 10 millimeters = 1 centimeter. Therefore, we can simplify 13 millimeters to 1 centimeter 3 millimeters. The answer, 19 centimeters 13 millimeters, then, becomes 20 centimeters 3 millimeters.

Example C: From 18 centimeters 5 millimeters subtract 6 centimeters 12 millimeters.

— Too small

18 centimeters	5 millimeters
− 6 centimeters	12 millimeters

Rename 1 centimeter as 10 millimeters.

17	15
1̶8̶ centimeters	5̶ millimeters
− 6 centimeters	12 millimeters
11 centimeters	3 millimeters

Example D: From 9 meters 2 millimeters subtract 5 meters 8 millimeters.

```
                              ← Too small
        9 meters     2 millimeters
       −5 meters     8 millimeters
       _____
```

Rename 1 meter to 100 centimeters. Then rename 1 centimeter to 10 millimeters.

```
   8            99            12
   9̶ meters    1̶0̶0̶ centimeters  2̶ millimeters
  −5 meters                      8 millimeters
  _____
   3 meters    99 centimeters    4 millimeters
```

▶ *Practice:* On your own paper, work each problem below.

1) 2 meters 55 centimeters + 3 meters 62 centimeters

2) 7 meters 9 centimeters + 6 centimeters

3) 5 centimeters 4 millimeters + 11 centimeters 8 millimeters

4) 12 centimeters 2 millimeters − 8 centimeters 5 millimeters

5) 23 centimeters 8 millimeters − 16 centimeters 9 millimeters

6) 8 meters 19 centimeters − 2 meters 25 centimeters

7) 4 meters 5 millimeters − 1 meter 7 millimeters

8) 26 meters 8 centimeters − 3 meters 8 millimeters

1.5 MULTIPLYING LINEAR MEASUREMENTS

Multiplication with units of measure is much the same as multiplication with whole numbers. First, multiply the numerical part of the measurement. Then write the units of measure with the product.

Example A: Multiply 6 meters 16 centimeters by 2.

$$
\begin{array}{r}
6 \text{ meters } 16 \text{ centimeters} \\
\times \qquad\qquad 2 \\
\hline
12 \text{ meters } 32 \text{ centimeters}
\end{array}
$$

Because
2 × 16 = 32
and
2 × 6 = 12

Example B: 3 × (6 centimeters 8 millimeters)

$$
\begin{array}{r}
6 \text{ centimeters } 8 \text{ millimeters} \\
\times \qquad\qquad\quad 3 \\
\hline
18 \text{ centimeters } 24 \text{ millimeters, or}
\end{array}
$$

20 centimeters 4 millimeters

Because
10 mm = 1 cm

24 millimeters = 2 centimeters and 4 millimeters. The answer, 18 centimeters 24 millimeters, then, becomes 20 centimeters 4 millimeters.

1.6 DIVIDING UNITS OF LINEAR MEASURE

There are two methods for dividing units of linear measure. The first method uses decimal notation.

Method One

Example: Divide 16 meters 8 centimeters by 6.

Step 1: Write 16 meters 8 centimeters as meters. Use a decimal. Hint: Use the ratio method.

$$\frac{16 \text{ meters}}{1 \text{ meter}} = \frac{8 \text{ centimeters}}{100 \text{ centimeters}}$$

Therefore, 16 meters 8 centimeters = $16\frac{8}{100}$ meters, or 16.08 meters.

Step 2: 16.08 meters divided by 6 equals 2.68 meters.

The answer, rounded to the nearest whole number, is 3 meters.

Method Two

In Method Two the larger units of measure are converted to the smaller units.

Example A: Divide 16 meters 8 centimeters by 6.

Step 1: Convert 16 meters 8 centimeters to centimeters. Multiply 16 by 100 because 1 meter = 100 centimeters. 16 meters = 1,600 centimeters.

Step 2: 16 meters 8 centimeters = 1,608 centimeters.

Step 3: Divide 1608 centimeters by 6.
The answer is 268 centimeters.

Step 4: Convert the answer back to the original units. 268 centimeters = 2 meters 68 centimeters, because 1 meter = 100 centimeters.

Note: If rounding is necessary in step 3, round to the nearest whole number.

Example B: Divide 18 meters 5 centimeters by 3.

Step 1: Convert 18 meters 5 centimeters to centimeters. 18 meters = 1800 centimeters.

Step 2: 18 meters 5 centimeters = 1805 centimeters.

Step 3: Divide 1805 by 3. The answer is 601.666... Rounded to the nearest whole number, this is 602, or 602 centimeters.

Step 4: Convert the answer back to the original units. 602 centimeters = 6 meters 2 centimeters.

▶ *Practice.* Multiply. Simplify the answers.

1) (6 meters 5 centimeters) × 7

2) (13 meters 55 centimeters) × 2

3) (2 meters 16 centimeters) × 20

4) (8 meters 5 millimeters) × 6

5) (6 centimeters 3 millimeters) × 7

6) (2 centimeters 7 millimeters) × 12

7) 5 × (5 meters 3 centimeters 5 millimeters)

8) 7 × (9 meters 80 centimeters 7 millimeters)

▶ Divide. Use Method One. Round answers to the nearest whole number.

9) (8 meters 12 centimeters) ÷ 4

10) (16 meters 15 centimeters) ÷ 5

11) (13 centimeters 3 millimeters) ÷ 2

12) (9 centimeters 4 millimeters) ÷ 3

13) (25 centimeters 8 millimeters) ÷ 6

14) (20 centimeters 5 millimeters) ÷ 10

▶ Divide. Use Method Two. Round answers if needed.

15) (5 meters 8 centimeters) ÷ 6

16) (12 meters 52 centimeters) ÷ 5

17) (5 centimeters 8 millimeters) ÷ 4

18) (4 centimeters 1 millimeter) ÷ 2

19) (11 meters 8 centimeters) ÷ 15

20) (7 centimeters 6 millimeters) ÷ 20

1.7 CALCULATING AREA

To find the area of a rectangle, multiply the length times the width. The units for length and width must be the same. The area will be in square units of measure.

Example A: What is the area of a rectangle 3 centimeters long and 2 centimeters wide?

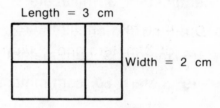

Length = 3 cm

Width = 2 cm

Area = length × width

= 3 cm × 2 cm

= 3 × 2 × cm × cm

= 6 × cm² or 6 cm²

The area is 6 square centimeters.

m × m	=	m²
cm × cm	=	cm²
mm × mm	=	mm²

Example B: What is the area of a rectangle 7 millimeters long and 6 millimeters wide?

7 mm

6 mm

Area = length × width

= 7 mm × 6 mm

= 7 × 6 × mm × mm

= 42 mm²

Example C: Find the area of a rectangle with a length of 8.5 cm and a width of 3.3 cm.

Area = length × width

= 8.5 cm × 3.3 cm

= 8.5 × 3.3 × cm × cm

= 28.05 cm²

Example D: Find the area of a rectangle with a length of 2 meters and a width of 55 centimeters. Remember that both units of measure must be the same. 1 meter = 100 centimeters.

Area = length × width

= 2 meters × 55 centimeters

or

= 200 centimeters × 55 centimeters

= 200 × 55 × cm × cm

= 11,000 cm²

Example E: Find the area of a rectangle with a length of 2 kilometers and a width of 752 meters.

If 1 kilometer = 1,000 meters
then 2 kilometers = 2,000 meters

Area = length × width

= 2 kilometers × 752 meters

or

= 2,000 meters × 752 meters

= 2,000 × 752 × m × m

= 1,504,000 m²

▶ **Practice.** Compute the area for each rectangle.

1) Length = 8 cm
   Width = 7.2 cm

2) Length = 3.5 m
   Width = 4 m

3) Length = 3.4 mm
   Width = 5.2 mm

4) Length = 2.6 m
   Width = 4.7 m

5) Length = .24 cm
   Width = 3 cm

6) Length = 11.75 m
   Width = 2.2 m

7) Length = 5.2 mm
   Width = 4 mm

8) Length = 13 m
   Width = 5.1 m

9) Length = 80 cm
   Width = 2 m

10) Length = 3.3 m
    Width = 4.3 m

11) Length = 8.4 mm
    Width = 0.2 mm

12) Length = 3 km
    Width = 47 m

13) Length = 100 cm
    Width = 3 m

14) Length = 10.75 m
    Width = 0.2 m

1.8 DIVIDING SQUARE UNITS OF MEASURE

Division is the opposite, or *INVERSE*, operation of multiplication. This can be seen by studying the simple multiplication fact, 2 × 6.

If 2 × 6 = 12
Then 12 ÷ 2 = 6
And 12 ÷ 6 = 2

The factors of 12 are 2 and 6. Dividing 12 by either factor will yield the other factor as the answer.

Example: Area = 3 cm × 2 cm

$$= 6 \text{ cm}^2$$

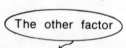
The other factor

Then 6 cm² ÷ 3 cm must equal 2 cm.

This division can also be expressed as a fraction.

$$\frac{6 \text{ cm}^2}{3 \text{ cm}} = \frac{6 \text{ cm} \times \cancel{cm}}{3 \; \cancel{cm}} = 2 \text{ cm}$$

1.9 EXAMPLES OF DIVISION PROBLEMS
Study these sample division problems.

1) $8 \text{ cm}^2 \div 2 \text{ cm} = 4 \text{ cm}$ or $\frac{8 \text{ cm}^2}{2 \text{ cm}} = \frac{8 \text{ cm} \times \cancel{\text{cm}}}{2 \cancel{\text{cm}}} = 4 \text{ cm}$

2) $12 \text{ m}^2 \div 6 \text{ m} = 2 \text{ m}$ or $\frac{12 \text{ m}^2}{6 \text{ m}} = \frac{12 \text{ m} \times \cancel{\text{m}}}{6 \cancel{\text{m}}} = 2 \text{ m}$

3) $\frac{12 \text{ cm}^2}{8 \text{ cm}} = \frac{\overset{3}{\cancel{12}} \text{ cm} \times \cancel{\text{cm}}}{\underset{2}{\cancel{8}} \cancel{\text{cm}}} = \frac{3}{2} \text{ cm}$ or $1\frac{1}{2} \text{ cm}$

See how this procedure can be applied to cubic units.

4) $\frac{8 \text{ cm}^3}{2 \text{ cm}^2} = \frac{8 \text{ cm} \times \cancel{\text{cm}} \times \cancel{\text{cm}}}{2 \cancel{\text{cm}} \times \cancel{\text{cm}}} = 4 \text{ cm}$

5) $\frac{32 \text{ m}^3}{8 \text{ m}} = \frac{32 \text{ m} \times \text{m} \times \cancel{\text{m}}}{8 \cancel{\text{m}}} = 4 \text{ m} \times \text{m} = 4 \text{ m}^2$

Of course, dividing by a simple whole number can be just as easy.

6) $\frac{\overset{9}{\cancel{18}} \text{ m}}{\underset{1}{\cancel{2}}} = 9 \text{ m}$ 7) $\frac{2 \text{ m}^2}{8} = \frac{1 \text{ m}^2}{4}$

▶ **Practice.** Divide. Simplify the answers.

1) $28 \text{ mm}^2 \div 7 \text{ mm} =$ 2) $35 \text{ cm}^2 \div 7 \text{ cm} =$

3) $\frac{16 \text{ m}^2}{32 \text{ m}} =$ 4) $\frac{20 \text{ cm}^3}{5 \text{ cm}} =$

5) $\frac{28 \text{ m}}{7 \text{ m}} =$ 6) $\frac{56 \text{ cm}^3}{28} =$

7) $22 \text{ mm}^2 \div 11 =$ 8) $39 \text{ cm}^2 \div 13 \text{ cm} =$

9) $12 \text{ cm}^2 \div 2 \text{ cm}^2 =$ 10) $56 \text{ m}^2 \div 14 \text{ m} =$

11) $25 \text{ cm}^2 \div 8 \text{ cm} =$ 12) $60 \text{ cm}^3 \div 12 \text{ cm} =$

13) $54 \text{ m}^3 \div 2 \text{ m} =$ 14) $18 \text{ mm}^3 \div 9 \text{ mm}^2 =$

1.10 MEASURING VOLUME

VOLUME is found by using the formula, length × width × height.

Example A: What is the volume of this rectangular box?

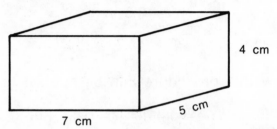

Remember, all edges that are parallel to each other have the same measurements.   To find the volume, use the formula:

Volume = length × width × height

You can multiply the factors in any order, as shown below.

$$V = 7 \text{ cm} \times 4 \text{ cm} \times 5 \text{ cm}$$
$$\text{or}$$
$$= 4 \text{ cm} \times 7 \text{ cm} \times 5 \text{ cm}$$
$$\text{or}$$
$$= 5 \text{ cm} \times 7 \text{ cm} \times 4 \text{ cm}$$

We will use the last combination to compute the volume.

$$V = 5 \text{ cm} \times 7 \text{ cm} \times 4 \text{ cm}$$
$$= 5 \times 7 \times 4 \times \text{cm} \times \text{cm} \times \text{cm}$$
$$= 35 \times 4 \text{ cm}^2 \times \text{cm}$$
$$= 140 \text{ cm}^3$$

The volume is 140 cubic centimeters.

Example B: Compute the volume for a large container with these measurements:

Length = 3.5 m   Width = 4.2 m   Height = 5.2 m

$$\text{Volume} = \text{length} \times \text{width} \times \text{height}$$

$$= 3.5 \text{ m} \times 4.2 \text{ m} \times 5.2 \text{ m}$$

$$= 14.7 \times 5.2 \text{ m}^3$$

$$= 76.44 \text{ m}^3$$

The volume is 76.44 cubic meters.

When you multiply cm × cm, the answer is in *SQUARE CENTIMETERS*. When you multiply cm × cm × cm, the answer is in *CUBIC CENTIMETERS*.

Example C: Compute the volume for a rectangular solid with dimensions of 5 mm, 8 mm, and 6 mm.

5 mm

8 mm 6 mm

$$V = 5 \text{ mm} \times 8 \text{ mm} \times 6 \text{ mm}$$

$$= 40 \times 6 \text{ mm}^3$$

$$= 240 \text{ mm}^3$$

The volume is 240 cubic millimeters.

► **Practice.**

1) A test tube box measures 8 centimeters by 9 centimeters by 12 centimeters. Compute the volume.

2) What is the volume of a stainless steel container whose length = 18 mm, width = 20 mm, and height = 10 mm?

3) Compute the volume for a laboratory supply cabinet that measures 1 m 2 cm by 5 m by 75 cm. Hint: Convert all meters to centimeters. 1 meter = 100 centimeters.

4) Compute the volume for this rectangular solid:

1 cm

4 cm

2 cm

1.11 MEASURING LIQUID VOLUME

The measurement of liquid is based on the cubic centimeter. A cubic centimeter is shown below:

One thousand cubic centimeters (cm^3) is a unit of measure named a *LITER*. You may see liter containers at the supermarket, especially in the soft drink section.

Sometimes we need to express cubic centimeters as liters. We can do this by dividing by 1,000.

Example A: Express 1,256 cm^3 as liters.

1,256 ÷ 1,000 = 1.256 liters

Moving the decimal point three places to the left is the same as dividing by 1,000.

Example B: Express 803 cm^3 as liters.

803 ÷ 1,000 = 0.803 liters

Or 0.803 L

The abbreviation for *liter* is the capital letter L.

▶ *Practice.* Express these cubic centimeters as liters.

1) 28,412 cm³   2) 35,162 cm³   3) 2,000 cm³

4) 4,160 cm³   5) 78,002 cm³   6) 15,000 cm³

7) 2,105 cm³   8) 98 cm³   9) 1,000 cm³

10) 706 cm³   11) 150 cm³   12) 37 cm³

13) 371 cm³   14) 7,113 cm³   15) 5 cm³

Notice how the metric prefixes are used with the liter:

Equivalents Using Liters
1 kiloliter = 1,000 liters
1 liter   = 100 centiliters
1 centiliter = 10 milliliters
1 liter   = 1,000 milliliters

One interesting equivalent is 1 liter = 1,000 milliliters. A milliliter is 1/1,000 of a liter, which happens to be a cubic centimeter. This may sound confusing, but there are two names for the same quantity.

Equivalents Using Cubic Centimeters
1 cubic centimeter = 1 milliliter
1,000 cubic centimeters = 1,000 milliliters
1,000 cubic centimeters = 1 liter

▶ **Practice.** Convert each of these measurements. Use the hint in parentheses.

1) 3 liters = _____ milliliters (times 1,000)

2) 5.5 liters = _____ milliliters (times 1,000)

3) 3,000 cubic centimeters = _____ liters (divide by 1,000)

4) 3,700 cm³ = _____ liters (divide by 1,000)

5) 0.72 liters = _____ milliliters (times 1,000)

1.12 MEASURING MASS

Picture yourself as an astronaut with super powers, bouncing through space, jumping from one planet to another. Your weight would change from one planet to another, because each planet would have a different gravitational pull on your body. You could jump higher on one planet than on another. Even though your weight would change from planet to planet, your mass (physical make-up) would remain the same. Mass and weight can be used interchangeably as long as we define our environment and mass with respect to the gravitational pull of earth.

The *GRAM* is a unit of mass (weight) that equals the mass of one cubic centimeter of water. This is about the same mass as a wooden match.

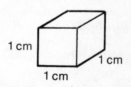

One cubic centimeter of water weighs one gram.

Equivalents of Units of Mass (Weight)

1 kilogram = 1,000 grams

1 gram   = 1,000 milligrams

1 gram   = 100 centigrams

1 centigram = 10 milligrams

If one cubic centimeter of water weighs one gram, then 1,000 cubic centimeters will weigh 1,000 grams, or one kilogram. Also, 1,000 cubic centimeters equals one liter. Therefore, one liter of water weighs one kilogram.

Your body weight might be measured in kilograms. However, the weight of a single hair from your head would be measured in milligrams. Centigrams are not used very often.

▶ *Practice.* Make these simple conversions.

1) 6 grams = ____ milligrams

2) 80,000 grams = ____ kilograms

3) 90 grams = ____ centigrams

4) 3,000 centigrams = ____ milligrams

▶ Express these weights in decimal notation.

5) 6 grams 5 milligrams

6) 18 grams 65 milligrams

7) 7 grams 16 milligrams

8) 24 grams 2 milligrams

1.13 TABLES OF METRIC EQUIVALENTS

Compare the three charts. Notice that each chart has a different root word. Notice that the same prefixes are used in all three charts.

Units of Length

1 kilometer	= 1,000 meters
1 meter	= 1,000 millimeters
1 meter	= 100 centimeters
1 centimeter	= 10 millimeters

Units of Volume

1 kiloliter	= 1,000 liters
1 liter	= 1,000 milliliters
1 liter	= 100 centiliters
1 centiliter	= 10 milliliters

Units of Mass (Weight)

1 kilogram	= 1,000 grams
1 gram	= 1,000 milligrams
1 gram	= 100 centigrams
1 centigram	= 10 milligrams

• CHAPTER REVIEW EXERCISES •

Vocabulary Words

meter	cubic	gram
kilometer	square	kilogram
centimeter	liter	centigram
millimeter	mass	milligram

Abbreviations

km = kilometer	kL = kiloliter	kg = kilogram
m = meter	L = liter	g = gram
cm = centimeter	cL = centiliter	cg = centigram
mm = millimeter	mL = milliliter	mg = milligram

▶ Copy these sentences onto your paper. Fill in the missing words.

1) Grams are used to measure _____, or weight.

2) You would use _____ to measure the height of a four-story building.

3) The length of an ant would be measured with _____.

4) The weight of a 19-inch television set would be measured in _____.

▶ Write *True* or *False* for each statement below.

5) One centimeter equals ten milliliters.

6) The volume of a rectangular solid with dimensions of 6 cm, 8 cm, and 5 cm is 240 cm^3.

7) Kilometers are used to measure the weight of meters.

8) Liters are used to measure liquid volume.

9) Cubic centimeters are used to measure weight.

10) The prefix *kilo-* means 100.

Computer Program

Many of the road signs you see today include distance measurements in metric units. This program will change kilometers to miles and miles to kilometers.

```
10 CLS
20 PRINT "METRIC/ENGLISH CONVERTER"
30 PRINT
40 PRINT
50 PRINT "PRESS:";TAB(10);"TO:"
60 PRINT "------";TAB(10);"---"
70 PRINT
80 PRINT
90 PRINT "1";TAB(10);"CHANGE MILES TO KILOMETERS"
100 PRINT
110 PRINT "2";TAB(10);"CHANGE KILOMETERS TO MILES"
120 PRINT
130 PRINT "3";TAB(10);"QUIT THE PROGRAM"
140 PRINT:PRINT
150 INPUT "WHAT IS YOUR CHOICE? ";MC$
160 IF MC$="" THEN 10
170 MC=VAL(MC$)
180 IF MC > 3 THEN 10
190 IF MC=3 THEN END
200 IF MC=2 THEN 280
210 CLS
220 PRINT "CHANGE MILES TO KILOMETERS"
230 PRINT:PRINT
240 INPUT "TYPE THE NUMBER OF MILES: ";M
250 PRINT:PRINT
260 PRINT M;" MILES EQUALS ";1.609 * M;" KILOMETERS."
270 GOTO 340
280 CLS
290 PRINT "CHANGE KILOMETERS TO MILES."
300 PRINT:PRINT
310 INPUT "TYPE THE NUMBER OF KILOMETERS: ";K
320 PRINT:PRINT
330 PRINT K;" KILOMETERS EQUALS ";.621 * K;" MILES."
340 PRINT:PRINT
350 INPUT "PRESS RETURN TO CONTINUE.";Z$
360 GOTO 10
```

Chapter 2

The Properties
of Matter

Chapter Goals:

1. To define the words *chemistry*, *matter*, *mass*, and *volume*.

2. To describe various objects by listing their properties.

3. To measure the volume of a liquid using a graduated cylinder.

4. To measure the volume of an object using the "displacement of water" method.

5. To measure the mass of different objects.

2.1 WHAT IS CHEMISTRY?

CHEMISTRY is the study of matter. *CHEMISTS*, people who study chemistry, describe what matter looks like, what it is made of, and how it changes. Chemists often work in laboratories where they have instruments to help them in their work.

Many different types of companies hire chemists to help find and test new products such as soaps, paints, foods, drugs, and fertilizers. Regardless of where they work, chemists must use skills and knowledge which they first gained from their chemistry courses in school.

A basic understanding of chemistry principles and facts can help the person who does not become a professional chemist. For example, the human body is made from matter, and uses other matter to stay alive. Knowing how your body works can help you plan good nutrition to keep your body healthy.

One common trait of all scientists is the desire to know about the world in which we live. They ask questions such as, "Why does sugar dissolve in water, while sand does not?" In other words, they have curiosity about the world. They want to know how and why things are the way they are. To answer their questions, they design and perform EXPERIMENTS to test their ideas. We will also use experiments to find out answers to our questions in chemistry.

2.2 WHAT IS MATTER?

To begin our study of chemistry, we need to define the word *matter*. A simple definition is that *MATTER* is anything that has mass and takes up space. Examples of matter are everywhere. Water, rocks, paper, steel, and air are all matter.

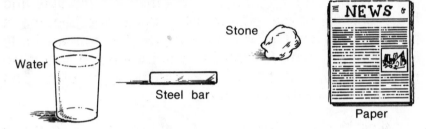

Water Steel bar Stone NEWS Paper

Examples of Matter

Each of these things takes up space, and each has mass. *MASS* is a measure of how much material is in an object.

2.3 MASS AND WEIGHT

Mass and *weight* are often used interchangeably to mean the same thing. Scientists, however, have different meanings for these two words. *MASS*, as we said, means how much material is in something. This amount never changes under normal conditions. *WEIGHT*, on the other hand, is a measure of how hard the earth's gravity pulls the object. We can measure weight with a bathroom scale.

Bathroom scale

The weight of an object can change by moving it to some other place. Astronauts who went to the moon *weighed* less on the moon than they did on the earth because the moon's mass is less than earth's. The moon's gravity, therefore, is also less than the earth's. While on the moon, the *mass* of the astronauts was the same as on earth.

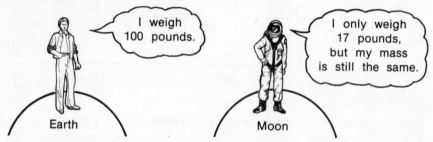

▶ *Practice.* In the table below are listed some of the members of the solar system along with their force of gravity compared to earth's. Calculate the weight of a 100-pound person on each of the different bodies. The first two examples are done for you. Write your answers on your own paper.

Planets/ Satellites	Force of Gravity Compared to Earth	Weight on Earth	Weight on This Planet	Method
Earth	1	100 lbs.	100 lbs.	1 × 100
Jupiter	2.54	100 lbs.	254 lbs.	2.54 × 100
Mars	.379	100 lbs.		
Saturn	1.07	100 lbs.		
Mercury	.378	100 lbs.		
Venus	.894	100 lbs.		
Moon	.165	100 lbs.		

2.4 WHAT ARE PROPERTIES?

If someone asked you to describe salt, you might say, "It is a white solid made of small individual grains." Each part of that description is a *PROPERTY* of salt. We could list the properties of salt as:

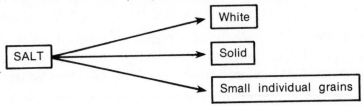

This description, although correct, is not enough to identify salt absolutely. The same properties are found in sugar. The description could be improved by adding another property, taste. Salt, then, is a white solid made of small, individual grains which have a "salty" taste. The sugar could be described as a white solid made of small, individual grains which have a sweet taste.

The chart below lists some of the more common properties which scientists use to describe matter.

Properties of Matter	
color	size
shape	mass
weight	taste
feel	smell

Scientists prefer to use some properties more than others. Mass and weight, for example, are more useful than color or shape. This is because mass and weight are easily *MEASURED*, and everyone can agree on the measurement. Another property that can be readily measured is size (length, width, and height).

Properties such as color are not measurable. Because of this, descriptions based on color can be misunderstood. To one person, a shade of blue might be described as "powder blue," while to another it would be "turquoise."

▶ **Practice.** On your paper, make a table like the one below. Then complete the table by listing as many properties as you can for each of the eight substances.

Substance	Properties
Milk	
Snow	
Charcoal	
Copper	
Soil	
Glass	
Water	
Air	

2.5 MEASURING MASS

MASS is defined as a measure of how much matter an object contains. In the metric system, mass uses the units known as *GRAMS*. To measure mass, we use a *SCALE*. There are many different kinds of scales, but the simplest kind often looks like the one below. It is sometimes referred to as a balance scale.

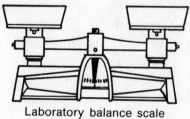

Laboratory balance scale

To use a scale of this type, the object whose mass is to be measured is placed in the pan on the left. When this is done, the balance arm tips to the left. Standard mass units (small brass cylinders with their mass stamped on them) are placed in the pan on the right, one at a time, until the scale is balanced. When balanced, the mass of the object on the left is equal to the sum of all the individual standard masses in the right pan. Look at the example below.

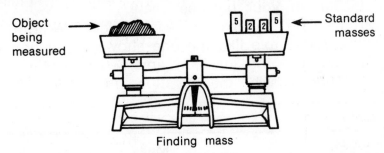

Finding mass

The mass of the object is equal to the sum of the masses on the right:

$$\text{Mass} = 5 \text{ g} + 5 \text{ g} + 2 \text{ g} + 2 \text{ g}$$

$$\text{Mass} = 14 \text{ g}$$

Being able to estimate mass is a valuable tool in chemistry. To do that, you need to know the mass of some common items.

INVESTIGATION 1

Purpose: To measure the mass of common objects using a balance scale.

A. Materials

Quantity:	Item Description:
1	Balance scale
1	Set of standard masses

Quantity:	Item Description:
1	Paper clip
1	Penny
1	Quarter
1	Nickel
1	Dime

B. Procedure
1) Set up your balance scale.
2) Measure the mass of each object. Write it on a data table like the one shown in Section C.
3) Measure the mass of any two other items.
4) Then answer the questions in Section D.

C. Data Table

Object	Mass
Paper clip	
Penny	
Quarter	
Nickel	
Dime	

D. Questions
1) What is the mass of one penny?
2) What would the mass of two dimes be?
3) What is the mass of one quarter plus one dime?
4) Find the mass of three dimes plus one quarter.
5) How much more mass does a quarter have than a penny?
6) What is the difference between the mass of a quarter and the mass of one nickel?

2.6 MEASURING THE MASS OF LIQUIDS

Finding the mass of a liquid presents a special problem. Balance scales were designed to hold solids, not liquids. In addition, some of the pans on balance scales are made of metal, which will begin to rust if liquids are placed in them.

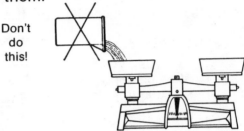

Don't do this!

Before we look at how to measure a liquid's mass, consider this similar problem. Suppose a boy weighs himself and finds his weight to be 100 pounds. Then he picks up his dog and weighs himself again. Of course, he will now weigh more than 100 pounds because of the extra weight of the dog.

100 lbs.

120 lbs.

The dog weighs 120 lbs. − 100 lbs., or 20 lbs.

The boy's weight with the dog is measured at 120 pounds. The dog, then, weighs:

120 pounds − 100 pounds = 20 pounds

Boy plus dog   Boy   Weight of dog

This same principle can be applied to mass. To measure a liquid's mass, first measure the mass of an empty *BEAKER*. A beaker is a plastic or glass cylinder similar to a measuring cup, which is used to hold liquids. Next, pour the liquid to be measured into the beaker and again measure the mass. Subtract the two masses. The answer will be the mass of just the liquid.

Mass of empty
beaker = 100 g

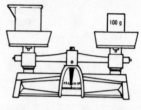

Mass of beaker
and liquid = 125 g

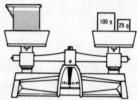

Mass of liquid = mass of beaker plus liquid – mass of beaker

125 g – 100 g

Mass of liquid = 25 g

▶ *Practice.* Write the answer to each question on your paper.

1) A beaker has a mass of 150 g. What is the mass of a liquid if the beaker plus liquid has a mass of 185 g?

2) A beaker has a mass of 125 g. What is the mass of a liquid if the beaker plus liquid has a mass of 163 g?

3) If the mass of a liquid is 35 g, and it is placed in a beaker having a mass of 75 g, what will be the mass of the beaker plus liquid?

2.7 MEASURING THE VOLUME OF LIQUIDS

We defined *matter* as anything that has mass and takes up space. *VOLUME* is a measure of how much space an object takes up. The volume of liquids is measured using a *GRADUATED CYLINDER*, a round tube like the one shown here:

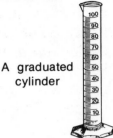

A graduated cylinder

The unit of volume in the metric system is the *LITER*. Usually, a liter is too large a unit, so the *MILLILITER* (1/1000th of a liter) is used instead. Remember that 1 milliliter has the same volume as 1 *CUBIC CENTIMETER*. Because of this, liquid volumes are often measured in cubic centimeters rather than milliliters.

There are many different sizes of graduated cylinders. The largest usually holds 1 liter, while more common sizes hold 100 milliliters, 50 milliliters, or 10 milliliters.

To measure the volume of a liquid, follow this procedure:

1. Pour the liquid into the graduated cylinder.
2. Hold the graduated cylinder up to your eye and sight across the top of the liquid.
3. Read the volume from the scale which is on the outside of the cylinder.
4. The level of the liquid usually has a rounded top. This rounded top is called a *MENISCUS*. Read the scale on the bottom of the curve as shown below.

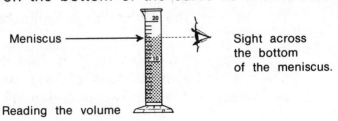

Meniscus

Sight across the bottom of the meniscus.

Reading the volume

Reading the scale on a graduated cylinder can be confusing. Look at the two cylinders below:

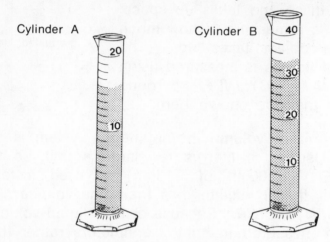

Cylinder A   Cylinder B

Cylinder A contains 16 mL, or 16 cc, of liquid. Cylinder B contains 34 mL, or 34 cc, of liquid. Here is how we determine the volume using the scales on the cylinders.

First, subtract the numbers on the two successive LONG lines. In cylinder A this would be 20 mL − 10 mL = 10 mL. Now, divide that answer by the number of spaces between two of the long lines. 10 mL divided by 10 = 1 mL per line. This means that for cylinder A, each line represents an increase of 1 mL from the line below. Since the liquid in cylinder A is 6 lines past the 10 mL mark, the volume is equal to:

$$10 \text{ mL} + (6 \times 1 \text{ mL}) = 16 \text{ mL}$$

For cylinder B, subtract: 30 mL − 20 mL = 10 mL. There are 5 spaces between each of the long lines. Therefore, we divide 10 mL by 5 to get 2 mL per line. The liquid in cylinder B is 2 lines past the 30 mL mark. The volume, then, is:

$$30 \text{ mL} + (2 \times 2 \text{ mL}) = 34 \text{ mL}$$

▶ **Practice**.

A. Read the volume of liquid in each cylinder.

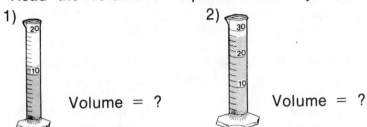

1)   Volume = ?

2)   Volume = ?

B. Draw two cylinders on your paper. Shade each one to show the given volume.

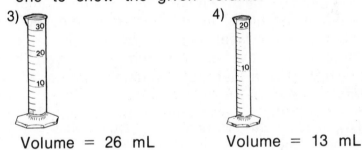

3)   Volume = 26 mL

4)   Volume = 13 mL

2.8   MEASURING THE VOLUME OF SOLID OBJECTS

Solid objects cannot be poured into graduated cylinders. However, we can measure the volume of these solids in two different ways, depending on the shape of the solid.

All solids can be described as having either a *REGULAR* shape or an *IRREGULAR* shape. Objects A and B below have a regular shape. Objects C and D have an irregular shape.

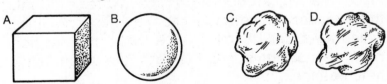

Regular shapes   Irregular shapes

When a solid has a regular shape, the volume can be found by using a formula. You have already learned the formula for rectangular objects. It is:

Volume = length × width × height

Solids that have shapes other than rectangular have different formulas for finding their volume.

Formulas for volume do not apply to irregularly-shaped objects. Instead, we use the *DISPLACEMENT OF WATER* method to find their volumes.

If a glass is filled with water and then something is placed in the glass, the level of the water will rise. In fact, the water level will rise by an amount equal to the volume of the object which was placed in the glass. Instead of a glass, a graduated cylinder is used for this method.

To measure the volume of a small object, follow this procedure:

1. Pour water into a graduated cylinder. Record the volume of the water.

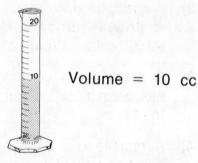

Volume = 10 cc

2. Place the object in the cylinder. The water level will then rise. Record this new volume.

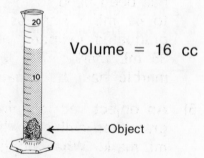

Volume = 16 cc

Object

3. Subtract the two volumes. The difference will be the volume of the object. $16 \text{ cc} - 10 \text{ cc} = 6 \text{ cc}$

There are two cautions to remember before using this method:

1. At the beginning you must have enough water in the graduated cylinder to completely cover the object when it is placed in the cylinder.

2. It does not matter how much water you start with as long as you have correctly done number 1 above.

▶ *Practice.* Find the volume for each of the following. Write the answers on your paper.

1) What is the volume of a rectangular object 10 cm long, 5 cm wide, and 2 cm high?

2) A graduated cylinder is filled to the 20 mL mark. A stone is dropped in, and the level of water rises to 32 mL. What is the volume of the stone?

3) A stone is placed in a graduated cylinder which has been filled to the 35 mL mark. The level rises to 42 mL. What is the volume of the stone?

4) A marble is placed in a graduated cylinder which has been filled to the 25 mL mark. The level rises to 41 mL. Another marble is placed in another graduated cylinder which has been filled to the 35 mL mark. The level rises to 52 mL. Which marble has the greater volume?

5) An object with a volume of 18 mL is placed in a graduated cylinder which has been filled to the 25 mL mark. What will the reading on the cylinder be?

INVESTIGATION 2

Purpose: To measure liquid volume and the volume of some common objects.

A. Materials

Quantity:	Item description:
1	Cap from tube of toothpaste
1	Styrofoam cup
1	Cap from bottle of mouthwash
1	Penny
1	Marble
1	Stone
1	Die from a game
1	Metric ruler
1	Graduated cylinder

B. Procedure

1) Fill the toothpaste cap with water. Then pour the liquid into a clean, dry graduated cylinder.

2) Measure the volume of water. Record this value on your data table.

3) Repeat steps 1 and 2 using the styrofoam cup and the mouthwash cap.

4) Measure the volume of the penny, marble, stone, and game die using the "displacement of water" technique. Record these values.

5) Measure the volume of the game die using your metric ruler. Make the measurements in millimeters. (Remember that a die is a regular object whose volume can be determined by the formula: Volume = length $\times$ width $\times$ height)

C. Data Table. Copy the table below on your paper. Write your data on your table.

Toothpaste cap	
Styrofoam cup	
Mouthwash cap	
Penny	
Marble	
Stone	
Die	

D. Questions. Write the answers on your paper.

1) Which object has the largest volume?

2) List the objects in order from smallest volume to largest volume.

3) You measured the volume of the die using two different methods. Do your answers agree?

4) Suppose you had a jar whose volume was greater than the capacity of your graduated cylinder. How could you measure the jar's volume with no additional materials?

5) Do you think it would be easy or hard to measure the volume of a very small item, for example, a pin, by the "displacement of water" method? Explain your answer.

6) Suggest a title for each of the columns in your data table.

SUMMARY OF CHAPTER 2

1) Chemistry is the study of matter.

2) Matter is anything that has mass and volume.

3) Mass measures how much material is in an object, while weight measures the earth's pull of gravity on the object.

4) Weight can change when moving from one place to another, but mass generally stays the same.

5) We use properties to describe matter.

6) Mass and volume are important properties of matter.

7) Mass is measured using a balance scale. The unit of mass is the gram.

8) Volume of liquids is measured using a graduated cylinder. The units of volume are the milliliter or the cubic centimeter.

9) When measuring the mass of liquids, do not pour the liquid into the pan on the balance. Measure the mass of an empty beaker. Then pour the liquid into the beaker and measure the mass again. Subtract these two figures to find the mass of the liquid.

10) The volume of regular objects can be calculated using formulas.

11) The volume of irregular objects is measured using the "displacement of water" technique.

• CHAPTER REVIEW EXERCISES •

Vocabulary Words

chemistry	mass	graduated cylinder
laboratory	volume	cubic centimeter
experiment	weight	meniscus
matter	property	milliliter
balance scale	gram	displacement of water

▶ Write the words that complete the sentences.

1) Liquid volume is measured with a ____ ____.

2) ____ is anything that has mass and takes up space.

3) When reading a graduated cylinder, you should sight across the ____.

4) ____ ____ ____ is the method of measuring the volume of irregular objects.

5) Scientists often work in ____.

▶ Write the letter of the best ending for each sentence.

6) Graduated cylinders are marked with:
   a) grams.   b) meters.   c) milliliters.

7) The volume of irregular objects is measured by:
   a) the displacement of water.
   b) a balance scale.
   c) a meter stick.

8) Ten cubic centimeters is the same as:
   a) 10 liters.   b) 10 milliliters.   c) 10 centiliters.

9) One liter is equal to:
   a) 1000 mL.   b) 100 mL.   c) 10 mL.

10) Volume is a measure of:
   a) an object's length.
   b) the amount of space an object takes up.
   c) an object's mass.

Calculator Practice

▶ Use your calculator to find the answers.

1) Find the volume of this figure by formula:

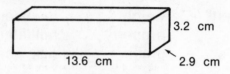

2) Perform the following operations:

a) 18.28 mL − 2.6 mL   b) 6.29 mL + 8.36 mL
c) 18.39 g × 2.3   d) 18.66 g − 3.77 g
e) 6.9 × 3.26 g   f) 3.96 mL − 2.01 mL
g) 16.62 g × 4.32   h) 332.6 mL + 129.6 mL

Computer Program

▶ Here is a program that will calculate volumes using the "displacement of water" technique. The *beginning volume* is the volume *before* you place the object in the cylinder. The *final volume* is the volume *after* you place the object in the cylinder.

```
10 CLS
20 PRINT "ENTER THE BEGINNING VOLUME";
30 INPUT BV
40 PRINT "ENTER THE FINAL VOLUME";
50 INPUT FV
60 LET V = FV − BV
70 CLS
80 PRINT "THE VOLUME IS: ";V
90 PRINT
100 PRINT "DO YOU WANT TO DO ANOTHER"
110 PRINT "PROBLEM?(Y/N)";
120 INPUT N$
130 IF N$ = "Y" THEN 10
140 CLS
150 END
```

Chapter 3

The Structure
of Matter

Chapter Goals:

1. To define the words *compound*, *atom*, *element*, *molecule*, *nucleus*, *proton*, *electron*, and *neutron*.

2. To draw diagrams of atoms showing protons, neutrons, and electrons.

3. To describe the meaning of atomic number and atomic mass number.

4. To calculate the number of protons, electrons, and neutrons in an element from the atomic number and atomic mass number.

3.1 INTRODUCTION

Pick up a newspaper and look at a picture. From a distance of a few feet it looks normal. Now, move the paper closer to your eyes. As it gets nearer, the picture is not as easily recognized. You may begin to see small black dots at a distance of one foot or less. If you have a magnifying glass, look at the picture through it. The dots become much larger, and the picture is more difficult to recognize.

It seems hard to believe, but thousands of dots in just the right place can make a true-to-life picture!

3.2 WHAT ARE MOLECULES?

Imagine looking at an individual grain of sugar. Certainly, it is not possible to see any small dots. Now imagine looking closer at the sugar grain. If it were possible to magnify the grain enough, we could see tiny particles which scientists call *MOLECULES*. Many different substances are made of molecules, but not all. Molecules are the smallest particles of a substance that still have the properties of that substance.

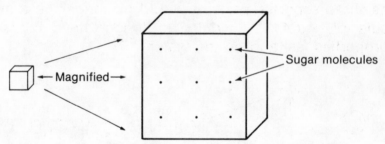

Sugar molecules

←Magnified→

Each molecule of sugar has the same properties as a grain of sugar. Each molecule would have a sweet taste.

Molecules are incredibly small. Millions of them could be placed side by side in the space of one centimeter. In addition to being small, they are constantly in motion. They even bump into each other occasionally.

Molecules are moving.

Even though the molecules are in motion, they tend to stay together because they attract each other.

Another substance that is made of molecules is ordinary water. Here is what a few molecules of water might look like if we could magnify them enough:

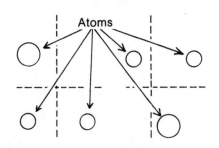

Four water molecules

As you can see, each molecule of water is made of three particles: one large particle and two smaller particles. Remember that each of these molecules of water has the same properties as a larger drop of water.

If one of these molecules were to be divided into its three separate parts, it would no longer be a molecule of water. It would no longer have the same properties as a drop of water.

Atoms

When these water molecules are separated into their parts, they are no longer water. The properties have changed.

When the water molecule is divided into its separate parts, the individual particles are called *ATOMS*. All substances are made of atoms. Substances that are made from atoms that are all the same are called *ELEMENTS*.

Substances that are made from two or more atoms of different elements are called *COMPOUNDS*.

INVESTIGATION 3

Purpose: To show that water is made from two different substances and is therefore a compound.

A. Materials

Quantity:	Item Description:
1	Beaker or wide-mouth jar
2	Dry cell batteries (hobby type 1.5 volt)
2	Pieces of copper wire, each about 50 cm long (preferably stiff)
1	Teaspoon of table salt

B. Procedure

1) Set up your materials as shown in this diagram:

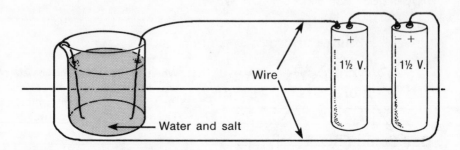

2) Fill the beaker with water (distilled if possible). Tap water will be all right if distilled water is not available.

3) Remove about 3 cm of insulation from the ends of the copper wire.

4) Position the wires in the beaker so that they are about 4 or 5 cm apart.

5) Connect the "+" terminal of one battery to the "−" terminal of the other battery.

6) Slowly dissolve the salt in the water by sprinkling a few grains at a time. Observe the ends of the wires in the beaker. Be sure to stir the solution each time you add salt.

7) When small bubbles begin to appear at the wire ends, you can stop adding the salt.

8) Observe the ends of both wires for about ten minutes.

C. Questions and Further Investigation

1) Which of the wires had more bubbles around it? Identify the wire by telling which terminal of the battery it was connected to (+ or −).

2) Write a description of your observations after you have waited for the full ten minutes.

3) What do you think would be the effect of adding another battery? Try it.

4) What do you think would happen if you added more salt? Try it.

D. Comments

1) The gas appearing at the wire connected to the − terminal is hydrogen gas. It is one of the elements contained in the compound water. The other gas, oxygen, appears at the positive wire.

2) The purpose of the salt is to make the water conduct electricity.

3.3 ELEMENTS AND COMPOUNDS

In the previous experiment, you saw that water could be broken down into simpler substances. In fact, water breaks down into two substances — hydrogen and oxygen. Both of these substances are gases. They are examples of a type of matter called *ELEMENTS*. Elements are the basic building blocks of all matter. There are 92 elements which we call *NATURAL* elements. Natural elements can be found in nature. There are a few more elements, but they are produced by scientists in specialized laboratories.

Many of the natural elements are quite common. Some of them are listed in the table below.

Some Common Elements	
Name:	**Used for or found in:**
Copper	Coins, frying pans, wire
Silver	Jewelry, photography
Carbon	"Lead" pencils, charcoal
Helium	Balloons
Nitrogen	The air which we breathe
Chlorine	Bleach
Aluminum	Airplanes
Neon	"Neon" signs
Gold	Jewelry
Mercury	Thermometers
Iron	Making steel

There is no way to know whether a substance is an element or compound simply by looking at it. It is necessary to do tests in a laboratory to determine which one it is.

When two or more elements combine chemically, they form a second class of substances called *COMPOUNDS*. When elements combine chemically to form compounds, the new substance can have completely different properties from the individual elements.

Table salt is a familiar compound which we eat every day. It is a compound formed from two elements, sodium and chlorine. Sodium is a solid, and chlorine is a gas that is poisonous. Combined chemically in the compound table salt, it is no longer poisonous.

Many familiar substances are examples of compounds. The table below lists some common compounds.

Some Common Compounds		
Name:	**Elements in this compound:**	**Use:**
Table salt	Sodium, chlorine	Cooking
Water	Hydrogen, oxygen	Drinking
Sugar	Carbon, hydrogen, oxygen	Cooking
Baking soda	Sodium, hydrogen, carbon, oxygen	Baking
Epsom salts	Magnesium, sulfur, oxygen	Medicine

3.4 ELEMENTS AND ATOMS

Just as compounds are made of molecules, molecules are made of atoms. Compounds can be broken down into separate atoms. ALL substances are made from atoms.

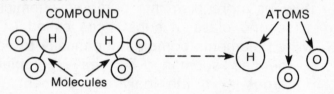

All atoms of the same element are alike. All atoms of oxygen would look the same. However, atoms of oxygen are different from atoms of all other elements.

3.5 THE STRUCTURE OF ATOMS

Even atoms are made of yet smaller particles. Scientists have made a *MODEL* of what they think atoms might look like. Since it is not possible to see things as small as atoms, scientists use models to help them describe how matter behaves. Scientists have had to change their model of atoms as new information is obtained.

Hydrogen

The simplest element in terms of its atoms is hydrogen. Hydrogen is a gas that will explode violently if it is ignited. At one time it was used to fill dirigibles. One of these dirigibles, the Hindenburg, burned and exploded when its hydrogen was ignited. Because of this, hydrogen is no longer used for this purpose.

This is a drawing of what an atom of hydrogen might look like:

An atom of hydrogen

In the central part of the atom, which we call the *NUCLEUS*, is a particle called a *PROTON*. Revolving around the proton is a smaller particle called an *ELECTRON*. To identify each of these particles, we can label them as in the drawing below.

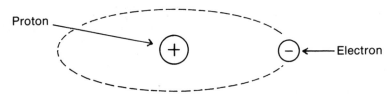

The symbol for a proton is a "+" sign. The symbol for an electron is a "−" sign. Sometimes the letter *P* is used for protons and *E* for electrons. We could use these letters in the drawing:

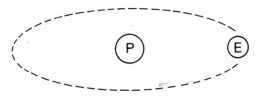

The broken line means that the electron travels around the proton just as a stone on a string can be twirled around your head. The electron is not tied to the proton, however. It does not fly off because it is attracted to the proton.

Helium

The next element is helium. Helium is also a gas that is often used in balloons. Putting helium into a balloon causes the balloon to rise, because the gas is lighter than air.

Helium atoms are different from hydrogen atoms because there are two protons and two electrons. An atom of helium might look like this:

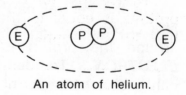

An atom of helium.

The nucleus now contains two protons, and there are two electrons revolving around them.

Lithium

You might be able to guess that the third element, lithium, has three protons and three electrons. A drawing of this atom is at the right.

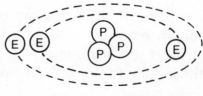

An atom of lithium.

3.6 A TABLE OF ELEMENTS

As we continue with the other elements, it becomes increasingly difficult to draw diagrams of them. For this reason, scientists rarely use these diagrams for any but the simplest of atoms. In the table below is a listing of the first ten elements. The number of protons and electrons is given for each element.

Elements with Number of Protons and Electrons		
Name	Number of Protons	Number of Electrons
Hydrogen	1	1
Helium	2	2
Lithium	3	3
Beryllium	4	4
Boron	5	5
Carbon	6	6
Nitrogen	7	7
Oxygen	8	8
Fluorine	9	9
Neon	10	10

Notice that for each element listed, the number of protons is equal to the number of electrons. This is true for all normal atoms. Study this rule:

Number of protons in an atom	=	Number of electrons in an atom

▶ *Practice.*

1) How many protons are in the element shown in the drawing?

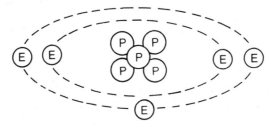

2) How many electrons are in the same element?

3) What is the name of this element? See page 54.

4) Draw a diagram of an element with 4 protons and 4 electrons.

5) Draw a diagram of the element carbon.

6) An atom of fluorine has 9 electrons. How many protons does it have?

7) Copy the following diagram. Place a *P* on each proton. Place an *E* on each electron.

3.7 THE ATOMIC NUMBER OF AN ELEMENT

Look again at the table on page 54. Notice that the difference between elements is in the number of protons and electrons that an atom has. Chemists describe elements with the *ATOMIC NUMBER*. The atomic number of an element is equal to the number of protons in the element.

The element hydrogen, for example, has an atomic number of 1 because it has 1 proton. The element oxygen, which has 8 protons, has an atomic number equal to 8. The element uranium has atomic number 92 because it has 92 protons in its nucleus. You can see the atomic numbers for each of the first ten elements in the table below.

Elements and Their Atomic Numbers			
Name	Atomic Number	Number of Protons	Number of Electrons
Hydrogen	1	1	1
Helium	2	2	2
Lithium	3	3	3
Beryllium	4	4	4
Boron	5	5	5
Carbon	6	6	6
Nitrogen	7	7	7
Oxygen	8	8	8
Fluorine	9	9	9
Neon	10	10	10

▶ **Practice.** The following table lists the next ten elements. Copy the table on your paper. Complete your table. Fill in the missing numbers.

Remember: 1. Number of protons = number of electrons

2. Atomic number = number of protons

Name	Atomic Number	Number of Protons	Number of Electrons
Sodium	11		
Magnesium		12	
Aluminum			13
Silicon	14		
Phosphorus			15
Sulfur		16	
Chlorine	17		
Argon		18	
Potassium	19		
Calcium			20

▶ Write the answers to these questions.

1) An element has an atomic number of 33. How many protons does it have?

2) An element has 26 protons. What is its atomic number?

3) An element has 46 electrons. How many protons does it have?

4) What is the name of the element with atomic number 14? Use the table above.

5) What is the name of the element that has 12 protons? Use the table above.

6) What is the name of the element that has 20 protons?

3.8 SOME ATOMS HAVE NEUTRONS

The atoms of some elements contain another particle. This particle is called a *NEUTRON*. A neutron is about the same size as a proton. The neutron is found in the nucleus of the atom with the protons. The neutron is identified with a letter "N" as in the following diagrams:

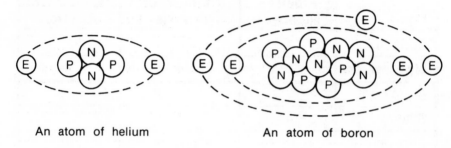

An atom of helium   An atom of boron

3.9 THE ATOMIC MASS OF AN ELEMENT

The mass of an electron is much less than the mass of a proton or neutron. In fact, the mass of a proton or neutron is about 1800 times the mass of an electron. Still, it would be impossible to measure the actual mass of a proton or neutron because they are much too small.

Chemists use the *ATOMIC MASS NUMBER* of an element instead of actually measuring the mass with a balance scale. The atomic mass number of an element is equal to the sum of the number of protons and neutrons in an atom of an element.

Atomic mass number = Number of protons + Number of neutrons

The following diagram is a drawing of an atom of beryllium, showing all of the major particles in an atom.

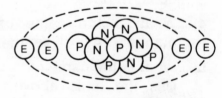

An atom of beryllium

Beryllium has 4 protons, 4 electrons, and 5 neutrons. The atomic number of beryllium is 4, since the atomic number is equal to the number of protons. The atomic mass number of beryllium is equal to 9, since there are 4 protons and 5 neutrons (4 + 5 = 9).

The table below is a summary of information for the first ten elements.

Information about the First Ten Elements					
Name	Atomic Number	Atomic Mass Number	Number of Protons	Number of Electrons	Number of Neutrons
Hydrogen	1	1	1	1	0
Helium	2	4	2	2	2
Lithium	3	7	3	3	4
Beryllium	4	9	4	4	5
Boron	5	11	5	5	6
Carbon	6	12	6	6	6
Nitrogen	7	14	7	7	7
Oxygen	8	16	8	8	8
Fluorine	9	19	9	9	10
Neon	10	20	10	10	10

If you know an element's atomic mass number and the number of protons in the element, you can determine the number of neutrons. Recall that:

Relationship 1:	Atomic mass number	=	Number of protons	+	Number of neutrons

Example for beryllium: 9 = 4 + 5

Relationship 1 can be turned around to give Relationship 2:

Relationship 2:	Number of neutrons	=	Atomic mass number	−	Number of protons

Example for beryllium: 5   =   9   −   4

A third relationship can be made in this way:

Relationship 3:	Number of protons	=	Atomic mass number	−	Number of neutrons

Example for beryllium: 4   =   9   −   5

You can use these three relationships to find all of the information about a particular element.

Remember: Atomic number = Number of protons

and

Number of protons = Number of electrons

For example, here is information for the element aluminum:

Name	Atomic Number	Atomic Mass Number	Number of Protons	Number of Neutrons	Number of Electrons
Aluminum	13	27	13	14	13

Here is how the information was determined:
1) Number of protons = atomic number. So, the number of protons for aluminum = 13.

2) Number of electrons = number of protons. The number of electrons for aluminum = 13.

3) Using relationship #2, the number of neutrons = atomic mass number − number of protons. The number of neutrons for aluminum = 27 − 13 = 14.

▶ **Practice.** Write the answers to these questions.

1) An element has an atomic number of 16. How many protons does it have?

2) An element has an atomic number of 9. How many electrons does it have?

3) An element has 15 protons and 16 neutrons. What is its atomic mass number?

4) An element has 28 electrons. How many protons does it have?

5) An element has 42 electrons. What is its atomic number? Why?

▶ Copy this table on your paper. Fill in the missing numbers. Use the three relationships.

Name	Atomic Number	Atomic Mass Number	Number of Protons	Number of Neutrons	Number of Electrons
Carbon		12		6	
Silver		108	47		
Silicon				14	14
Calcium	20	40			
Iodine	53	127			
Chlorine	17			18	
Sulfur		32			16
Potassium		39	19		

SUMMARY OF CHAPTER 3

1) Molecules are the smallest particles of compounds that still have the properties of the compound.

2) Molecules are in constant motion.

3) The smallest particles of elements that still have the properties of the element are called atoms.

4) Atoms are made of protons, neutrons, and electrons.

5) Protons and neutrons are located in the nucleus of the atom.

6) Electrons revolve around the nucleus of an atom.

7) There are 92 natural elements.

8) In a normal atom, the number of protons is equal to the number of electrons.

9) The atomic number of an element is equal to the number of protons.

10) The atomic mass number of an element is equal to the number of protons plus the number of neutrons.

The structure of an atom

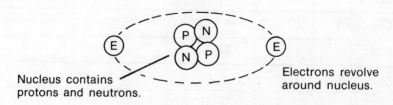

Nucleus contains protons and neutrons.

Electrons revolve around nucleus.

Number of protons = Number of electrons

• CHAPTER REVIEW EXERCISES •

Vocabulary Words

molecule	compound	atomic number
atom	model	electron
element	nucleus	atomic mass number
proton	neutron	

► Copy the sentences. Fill in the missing words.

1) The two particles found in the nucleus of an atom are the ＿＿ and ＿＿.

2) Two or more elements combined chemically are called a ＿＿.

3) There are ＿＿ natural elements.

4) The symbol for ＿＿ is "E."

5) The ＿＿ ＿＿ is equal to the number of protons.

► Write the letter of the best ending for each sentence.

6) The particle identified by an "N" is a
   a) proton.   b) electron.   c) neutron.

7) All substances are made from
   a) atoms.   b) models.   c) molecules.

8) The central part of an atom is called the
   a) element.   b) core.   c) nucleus.

9) The atomic mass number equals the number of protons plus the number of
   a) electrons.   b) neutrons.   c) atoms.

10) In a single atom, there is the same number of
   a) protons and neutrons.
   b) electrons and neutrons.
   c) protons and electrons.

Computer Program

▶ Here is a program that will calculate the information for any element. You must type in the atomic number and the atomic mass for the element. The program will then print all the information for that particular element.

```
10 CLS
20 PRINT "THIS PROGRAM WILL CALCULATE"
30 PRINT "THE NUMBER OF NEUTRONS FOR"
40 PRINT "AN ELEMENT.  YOU MUST TYPE"
50 PRINT "THE ATOMIC NUMBER AND THE"
60 PRINT "ATOMIC MASS."
70 PRINT
80 PRINT
90 PRINT "WHAT IS THE ATOMIC NUMBER  ";
100 INPUT AN
110 PRINT
120 PRINT
130 PRINT "WHAT IS THE ATOMIC MASS  ";
140 INPUT AM
150 CLS
160 PRINT "ATOMIC NUMBER:    ";AN
170 PRINT
180 PRINT "ATOMIC MASS:    ";AM
190 PRINT
200 PRINT "PROTONS:    ";AN
210 PRINT
220 PRINT "ELECTRONS:    ";AN
230 PRINT
240 PRINT "NEUTRONS:    ";AM-AN
250 PRINT
260 PRINT
270 PRINT "DO YOU WANT TO TRY ANOTHER? (Y/N)";
280 INPUT R$
290 IF R$="Y" THEN 10
300 CLS
310 END
```

Chapter **4**

Classifying Elements

Chapter Goals:

1. Name the symbols for 40 different elements.

2. Describe the information found in a periodic table.

3. Use a periodic table to obtain information about an element.

4. Describe an element as either a metal, a non-metal, or a noble gas.

5. Calculate the density of a substance.

4.1 INTRODUCTION

Scientists have identified 92 natural elements. Many of these elements have similar properties. Silver and aluminum, for example, are both shiny metals with a grayish color.

Even though many of their properties are the same, there are some that are different. Aluminum is light in weight and therefore good for use in aircraft. Silver, on the other hand, is too heavy for use in airplanes. Silver is also too expensive for this use.

4.2 SYMBOLS

Each of the elements has a *SYMBOL* which chemists use. A symbol is a shortened form of the name of an element. It is similar to an abbreviation.

Many English words have more than one abbreviation. For example, the state of Pennsylvania can be abbreviated as Pa., PA, Penn., or Penna. Symbols, however, must always be written in the same way.

The rules for writing symbols are:

1) All symbols have either one or two letters. No symbol has more than two letters.

2) The first letter of a symbol is a capital letter.

3) If a symbol has a second letter, the second letter is a lower-case letter.

4) There is no period at the end of a symbol.

The following table lists five elements along with the correct symbol.

Element Name	Correct Symbol	What is wrong with these symbols?
helium	He	HE
bromine	Br	BR
calcium	Ca	CA
chlorine	Cl	CL
zinc	Zn	ZN

In each of the above five examples, the symbol in the last column is incorrect because the second letter is a small *capital* letter. It should be a lower-case letter.

Here is a list showing the correct form of the lower case letters:

a or α	g	m	s	y
b	h	n	t	z
c	i	o	u	
d	j	p	v	
e	k	q	w	
f	l	r	x	

▶ **Practice.** Each of the following are incorrect symbols for elements. State what rule makes each one wrong. Some may break more than one rule.

1) Hyd
2) NA
3) mG
4) Ba.
5) n
6) Mag
7) HE
8) BE
9) ca
10) Hel
11) Oxy
12) aL

4.3 THE ELEMENTS AND THEIR SYMBOLS

There are 92 natural elements. Each of those elements has its own special symbol. Many of those 92 elements are rare, while others are quite common. Forty of the elements are common enough that they should be memorized. The 40 elements can be divided into four different groups based on how their symbols are formed.

Symbols using the first letter of the element name:

Element Name	Element Symbol		Element Name	Element Symbol
1. hydrogen	H		6. fluorine	F
2. boron	B		7. phosphorus	P
3. carbon	C		8. sulfur	S
4. nitrogen	N		9. iodine	I
5. oxygen	O		10. uranium	U

Symbols using the first two letters of the element name:

Element Name	Element Symbol	Element Name	Element Symbol
11. helium	He	17. calcium	Ca
12. lithium	Li	18. cobalt	Co
13. neon	Ne	19. bromine	Br
14. aluminum	Al	20. barium	Ba
15. silicon	Si	21. radium	Ra
16. argon	Ar		

Symbols using the first letter and one other letter from the element name:

Element Name	Element Symbol	Element Name	Element Symbol
22. magnesium	Mg	26. plutonium	Pu
23. chlorine	Cl	27. zinc	Zn
24. chromium	Cr	28. strontium	Sr
25. manganese	Mn	29. platinum	Pt

Symbols using letters not found in the name:

Element Name	Element Symbol	Element Name	Element Symbol
30. sodium	Na	36. gold	Au
31. potassium	K	37. mercury	Hg
32. iron	Fe	38. lead	Pb
33. silver	Ag	39. antimony	Sb
34. tin	Sn	40. copper	Cu
35. tungsten	W		

Why are the symbols in the last list so different from the names of the elements? These symbols are usually derived from the *LATIN* names for the elements. For example, the Latin word for *sodium* is "natrium." From "natrium" we get the symbol Na.

▶ *Practice.* Study the spelling of each element. Memorize the elements and their symbols. Use these hints to help you memorize the symbols:

1) Make a set of flash cards from index cards. Write the name of the element on one side. Write the symbol on the other side. Go through the cards one at a time and quiz yourself.

2) Memorize five symbols at a time until you can correctly identify all 40.

3) Have a friend say the name of an element and you give the symbol.

▶ Number your paper from 1 to 10, and then:

Write the symbol for each element:

1) helium
2) silver
3) gold
4) chlorine
5) calcium

Write the name for each of these symbols:

6) Hg
7) Ne
8) U
9) O
10) P

4.4 THE PERIODIC TABLE

For many years scientists noticed that some elements were similar to others. They tried different methods of arranging the elements in a table which would show these similarities. In the mid 1800's, a Russian chemist named Mendeleev was successful in designing the first chart which would show some of the similarities between the elements. That original chart was not perfect, and many elements did not seem to fit properly.

The modern version of Mendeleev's chart is called the *PERIODIC TABLE*. The name, "periodic," suggests that the properties of the elements are periodic — in other words, the properties tend to repeat.

The periodic table consists of rows and columns of boxes in which the elements are listed in atomic number order. The following table is a partial table containing the first 20 elements.

Periodic Table of the First 20 Elements

1 H							2 He
3 Li	4 Be	5 B	6 C	7 N	8 O	9 F	10 Ne
11 Na	12 Mg	13 Al	14 Si	15 P	16 S	17 Cl	18 Ar
19 K	20 Ca						

Families

In the periodic table the elements that are together in a column have similar properties. The properties that are similar are usually *chemical* properties. In other words, they react with other matter in the same way. Being in the same column or *FAMILY* of elements does not mean that they all look alike.

Lithium, sodium, and potassium are solids that react violently when placed in water. Fluorine and chlorine are both poisonous gases. Helium, neon, and argon are all gases which seldom combine or react with any other matter. Each of these groups of elements are in the same family, and they share similar properties.

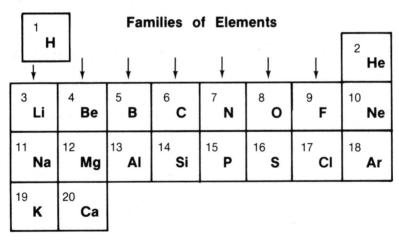

Families of Elements

Basic Information

Periodic tables can be simple and contain only the symbols for the elements. Or they can be quite complicated and contain much information about each of the elements. Most periodic tables have at least the following information:

1) The symbol for each element.
2) The atomic number for each element.
3) The atomic mass for each element.

Here is an example of how the information for hydrogen is shown on a periodic table:

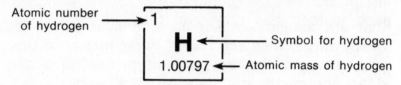

Atomic number of hydrogen

Symbol for hydrogen

Atomic mass of hydrogen

Notice that the atomic mass of hydrogen is not listed as a whole number. Most elements, in fact, have atomic masses listed in the periodic table as numbers with a decimal. The reason for this is that each element can be found in slightly different forms which are called *ISOTOPES*. An isotope of an element has the same number of protons and electrons as the original element, but has a different number of neutrons in the nucleus. The isotopes have the same properties as the element.

Hydrogen has three isotopes called *ordinary hydrogen*, *deuterium*, and *tritium*. An atom of each of these different forms is diagrammed below:

Hydrogen   Deuterium   Tritium

As you can see, the only difference between these three isotopes is the number of neutrons in the nucleus.

Most isotopes of elements do not have special names as the isotopes of hydrogen do. They are identified instead by their atomic mass. Look at the chart below.

Isotope	Atomic Number	Atomic Mass
hydrogen	1	1
deuterium	1	2
tritium	1	3

The atomic masses are different because of the additional neutrons in deuterium and tritium. The atomic number for each, however, is the same because each isotope has only one proton.

Because of the existence of isotopes, the atomic mass shown in the periodic table is an *average* mass for all of the isotopes of an element. The average depends on the number of isotopes and on the amount of each isotope found in nature.

The following table is a partial periodic table of the first 20 elements, with each element's atomic number and average atomic mass.

1 **H** 1.007																	2 **He** 4.002
3 **Li** 6.939	4 **Be** 9.012									5 **B** 10.811	6 **C** 12.011	7 **N** 14.006	8 **O** 15.999	9 **F** 18.998	10 **Ne** 20.183
11 **Na** 22.989	12 **Mg** 24.312									13 **Al** 26.981	14 **Si** 28.086	15 **P** 30.973	16 **S** 32.064	17 **Cl** 35.453	18 **Ar** 39.948
19 **K** 39.102	20 **Ca** 40.08														

As more elements are included in the periodic table, it becomes more difficult to maintain order. In fact, the table must be widened to fit all the elements into each of the families. The table on page 74 is a Periodic Table of the first 88 natural elements.

4.5 PERIODIC TABLE OF THE FIRST 88 NATURAL ELEMENTS

For the name of each element, see the Appendix, page 261.

1	2	3	4	5	6	7	8	9	10	11	12	13	14	15	16	17	INERT GASES
1 H 1.00797																	2 He 4.0026
3 Li 6.939	4 Be 9.0122											5 B 10.811	6 C 12.01115	7 N 14.0067	8 O 15.9994	9 F 18.9984	10 Ne 20.183
11 Na 22.9898	12 Mg 24.312											13 Al 26.9815	14 Si 28.086	15 P 30.9738	16 S 32.064	17 Cl 35.453	18 Ar 39.948
19 K 39.102	20 Ca 40.08	21 Sc 44.956	22 Ti 47.90	23 V 50.942	24 Cr 51.996	25 Mn 54.9380	26 Fe 55.847	27 Co 58.9332	28 Ni 58.71	29 Cu 63.54	30 Zn 65.37	31 Ga 69.72	32 Ge 72.59	33 As 74.9216	34 Se 78.96	35 Br 79.909	36 Kr 83.80
37 Rb 85.47	38 Sr 87.62	39 Y 88.905	40 Zr 91.22	41 Nb 92.906	42 Mo 95.94	43 Tc 99	44 Ru 101.07	45 Rh 102.905	46 Pd 106.4	47 Ag 107.870	48 Cd 112.40	49 In 114.82	50 Sn 118.69	51 Sb 121.75	52 Te 127.60	53 I 126.9044	54 Xe 131.30
55 Cs 132.905	56 Ba 137.34	57 La 138.91 ←	72 Hf 178.49	73 Ta 180.948	74 W 183.85	75 Re 186.2	76 Os 190.2	77 Ir 192.2	78 Pt 195.09	79 Au 196.967	80 Hg 200.59	81 Tl 204.37	82 Pb 207.19	83 Bi 208.980	84 Po 210	85 At 210	86 Rn 222
87 Fr 223	88 Ra 226																

☐ Metals
■ Non-Metals

The elements numbered 57 to 71 are called "rare-earth elements." We know very little about these elements. Only number 57 is included in this chart.

▶ **Practice.** Use the periodic table to answer the following questions.

1) What is the atomic number of helium?

2) What is the atomic mass of aluminum?

3) What is the symbol for carbon?

4) What element has an atomic number of 11?

5) What element has an atomic mass of 18.9984?

6) What is the atomic number for Mg?

7) What is the atomic mass of Ar?

8) What is the atomic number of oxygen?

9) What is the atomic mass of boron?

10) What happens to the atomic mass of an element as the atomic number increases?

11) What element is between carbon and oxygen?

12) What element is in the same family as chlorine?

13) What is the sum of the masses of the element oxygen and the element hydrogen?

14) What is the difference in the masses of the elements chlorine and fluorine? Subtract.

15) What is the difference in the masses of the elements sodium and lithium? Subtract.

4.6 SOME ELEMENTS ARE METALS

Nearly 80% of all the elements can be described as metals. Look back at the periodic table to see which elements are classified as metals.

All metals have similar properties. The table below lists some common properties of the metals.

Metals
Luster (shine) .... usually shiny.
Malleability....... can be hammered into thin sheets.
Ductility .... can be drawn into fine wires.
Conductivity ..... electricity and heat can travel through them easily.

Metals are important for many different uses. Gold, silver, and platinum are used in making jewelry. Copper and zinc are used in manufacturing coins. Mercury is used in thermometers and thermostats. Without metals our lives would be quite different!

 Jewelry

 Coins

Thermometer

Metals are usually solid at normal room temperature. (Mercury is one exception.) By heating metals, we can change them into liquids. By mixing liquid metals together and allowing them to cool and harden, we can make *ALLOYS*. Alloys are mixtures of metals. Alloys have a combination of the properties of the different metals.

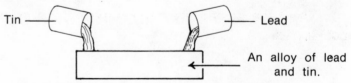

Tin

Lead

An alloy of lead and tin.

Some of the more important alloys of metals are listed in the table below.

Metal Alloys	
Alloy Name	**Made From**
solder..............	lead (Pb), tin (Sn)
bronze	copper (Cu), tin (Sn)
stainless steel	iron (Fe), chromium (Cr)
brass	copper (Cu), zinc (Zn), tin (Sn)
alnico..............	aluminum (Al), nickel (Ni), cobalt (Co), iron (Fe)

How are these alloys used? Solder is used for electrical connections. Bronze is used for artwork, such as plaques. Stainless steel is used for eating utensils. Brass is used for locks and plumbing. Alnico is used in magnets.

4.7 THE DENSITY OF ELEMENTS

You have probably heard the famous riddle, "Which weighs more, a pound of feathers or a pound of lead?" At first, many people say, "A pound of lead." But the answer is that they weigh the same — 1 pound each. This riddle illustrates another property of objects known as *DENSITY*. The density of lead is much greater than the density of feathers. A pound of lead would be a small cube, while a pound of feathers would be much larger. Another way of saying this is that the lead has more matter packed in a smaller space than feathers have.

The idea of density can be expressed in mathematical terms.

This is the formula for density:

DENSITY = MASS ÷ VOLUME

Suppose the mass of an object is measured as 30 grams, and its volume is measured as 15 cubic centimeters. Then the density would be calculated as:

DENSITY = MASS ÷ VOLUME

DENSITY = 30 grams ÷ 15 cubic centimeters

DENSITY = 2 grams/cubic centimeter

The density was found by dividing the mass of 30 grams by the volume of 15 cubic centimeters. Note that the density has the units of grams/cubic centimeters. This means that 1 cubic centimeter of this substance has a mass of 2 grams.

The density of any particular substance is always the same, regardless of the size of the object being measured. In the following Investigation, you will calculate the density of different substances.

INVESTIGATION 4

Purpose: To measure the density of some common materials.

A. Materials

Quantity:	Item Description:
2	Glass marbles
2	Rubber erasers
2	Pieces of chalk
2	Pieces of modeling clay (any size)
1	Balance scale
1	Graduated cylinder

B. Procedure

1) Using the graduated cylinder, measure the volume of each of the samples.

2) Using the balance scale, measure the mass of each of the samples.

3) Calculate the density for each of the objects.

4) Record your data in a data table.

C. Data Table. Copy the table below on your paper. Write your data in your table.

Item	Volume (cc)	Mass (g)	Density (g/cc)

D. Questions and Interpretation

1) Which object has the highest density?

2) Which object has the smallest volume?

3) Which object has the greatest mass?

4) List the objects in order from highest density to lowest density.

5) Does the object with the highest density also have the largest mass?

6) Do both of the marbles have the same density?

7) What is the density of a glass marble?

8) What is the density of rubber?

4.8 SOME ELEMENTS ARE NON-METALS

Look back again at the periodic table and find the elements known as non-metals. The non-metals are located on the right side of the table. Many of the non-metals are gases. Some of them are powders. This table shows some of the more common non-metals.

Non-Metals		
Name	**Symbol**	**Description**
sulfur .... S ......yellow powder *(Used in medicine, gunpowder)*		
oxygen .... O .....colorless, odorless gas *(Used in respiration)*		
chlorine.............. Cl .....greenish gas *(Used for bleaching)*		
iodine................ I ......gray-black solid *(Used for medicine)*		
carbon .... C .....many different forms, *(Diamonds, "lead" pencils)* mostly black		

4.9 THE NOBLE GASES

The six elements listed in the last column of the periodic table are called the *NOBLE GASES*. They are all gases and are inert. *INERT* means that they do not react or combine with any other elements under ordinary conditions. Many of the noble gases are used in making signs that light up with various colors when electricity is connected to them. These are the familiar "neon" signs.

SUMMARY OF CHAPTER 4

1) All elements have a symbol which is an abbreviation for its name.

2) The elements can be listed on a Periodic Table.

3) The elements in a column of the Periodic Table are called a *FAMILY*.

4) The Periodic Table lists the elements' symbols, atomic numbers, and atomic masses.

5) The elements of the Periodic Table are listed in order of increasing atomic number.

6) Most elements have isotopes, which are different forms of the same element.

7) An isotope of an element has the same number of protons and electrons as the original element. However, it has a different number of neutrons.

8) The atomic mass of an element is listed as an average mass of the various isotopes which exist in nature.

9) Elements are classified as either metals, non-metals, or noble gases.

10) The noble gases are also called *inert* because they do not ordinarily combine with any other elements.

11) The density of an object is its mass per unit volume.

12) To calculate density, the mass is divided by the volume. In the metric system, this is usually grams divided by cubic centimeters.

13) Metals can be melted and mixed to form alloys.

• CHAPTER REVIEW EXERCISES •

Vocabulary Words

symbol	isotope	tritium
periodic table	deuterium	metal
family	non-metal	noble gas
alloy	density	inert

▶ Write the word that completes each sentence.

1) The symbol Fe stands for ____.

2) A mixture of two or more metals is called an ____.

3) ____ is measured in g/cc.

4) Most of the elements are classified as ____.

5) Deuterium is an ____ of hydrogen.

▶ Write the letter of the best ending for each sentence.

6) A mass of 5 grams that has a volume of 2 cubic centimeters has a density of:
   a) 2.5 g/cc   b) 2.5 cc/g
   c) 10 g/cc   d) 10 cc/g

7) The atomic mass of hydrogen is:
   a) 1   b) 2   c) 3   d) 1.00797

8) An element which will not react with other elements is called:
   a) a metal   b) dense   c) inert

9) An element that is in the same family as fluorine is:
   a) lithium.   b) oxygen.
   c) sulfur.   d) chlorine.

10) The "noble gases" are found in the periodic table at the:
    a) right. b) left. c) top. d) bottom.

Calculator Practice

▶ Find the density for each of the following. Use your calculator.

1) 8.92 g ÷ 3.149 cc   2) 2382 g ÷ 542 cc
3) 183.2 g ÷ 42.1 cc   4) 823.946 g ÷ 282.3 cc
5) 19.001 g ÷ 14.2 cc   6) 66.0003 g ÷ 14 cc
7) 88 g ÷ 22.2 cc   8) 123 g ÷ 18 cc
9) .011 g ÷ .0001 cc   10) 4.2 g ÷ .989 cc
11) 3.2 g ÷ .899 cc   12) 45.3 g ÷ 13.23 cc

Computer Program

▶ Here is a program that will calculate the density of an object. You must type in the object's mass and its volume.

```
10 CLS
20 PRINT "PROGRAM TO CALCULATE DENSITIES"
30 PRINT:PRINT
40 PRINT "WHAT IS THE OBJECT'S MASS ";
50 INPUT M
60 PRINT:PRINT
70 PRINT "WHAT IS THE OBJECT'S VOLUME ";
80 INPUT V
90 LET D=M/V
100 PRINT:PRINT
110 PRINT "THE DENSITY = ";D
120 PRINT:PRINT
130 PRINT "DO YOU WANT TO DO ANOTHER? (Y/N) ";
140 INPUT R$
150 IF R$="Y" THEN 10
160 CLS
170 END
```

▶ Use the computer program to find the density of these two objects:

1) Mass = 38.2   2) Mass = 382.634
    Volume = 12.13   Volume = 76.39

Chapter 5

Compounds

Chapter Goals:

1. To explain how compounds are formed.

2. To describe the information contained in a formula.

3. To define the word *subscript*.

4. To list the elements and the number of atoms of each element in a compound.

5. To define the word *radical*.

6. To list the names of five radicals.

7. To name compounds containing only two elements.

8. To name compounds containing more than two elements.

5.1 WHAT ARE COMPOUNDS?

Although many natural substances are made of elements, the majority are made of *COMPOUNDS*. A compound is a chemical combination of two or more elements. The elements in a compound cannot simply be mixed together to form a compound. They must be combined chemically.

5.2 HOW ARE COMPOUNDS FORMED?

Elements are made from atoms. Atoms contain protons, neutrons, and electrons. The electrons can be visualized as revolving around the nucleus, which contains the protons and neutrons.

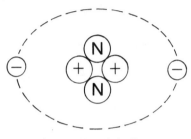

An atom of helium

The electrons in an atom form into shells around the nucleus. Each of these shells has a maximum number of electrons that it can hold. These shells can be thought of in the same way as the layers of an onion. An onion is made of layers which can be peeled away.

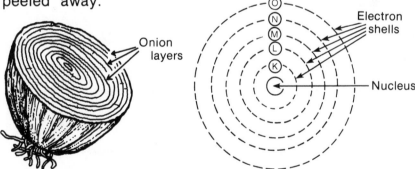

When atoms of different elements combine chemically to form compounds, they do so by sharing, lending, or borrowing electrons that are in their outermost shell. Atoms share, borrow, or lend electrons because by doing so, they form a more stable structure. The atoms are more stable when their shells contain the proper number of electrons.

Each shell has a letter name, as shown in this table:

Electron Shells in the Atom	
Shell Name	**Number of Electrons When Filled**
K	2
L	8
M	18
N	32
O	32

A familiar compound is table salt. Table salt is made from sodium and chlorine atoms. The outer shell of sodium has only 1 electron. The outer shell of chlorine has 7 electrons.

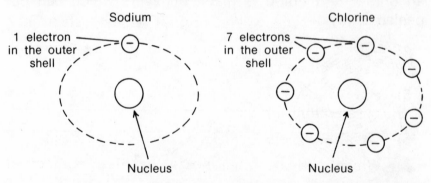

The atom of sodium will be more stable if it gives up the one electron in the outer shell. The chlorine will be more stable if it gains or borrows one electron.

The sodium atom lends its electron to chlorine, and, by doing so, both atoms become more stable. However, because the sodium atom now has more protons than electrons, it is said to have a "+" charge.

This means it has more protons than electrons. The chlorine atom now has more electrons than protons. We say that it has a "−" charge. When an atom has a charge, it is called an *ION*.

The atoms of sodium and chlorine are attracted to each other because all + charges will attract all − charges, just as the opposite poles of a magnet attract each other. This attraction force is what keeps the atoms together when they form a compound.

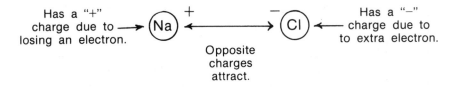

Has a "+" charge due to losing an electron.

Opposite charges attract.

Has a "−" charge due to to extra electron.

▶ *Practice.*

1) What is the definition of a compound?

2) What two particles are in the nucleus of the atom?

3) Why do atoms share, borrow, or lend electrons?

4) How many electrons are there in the outer shell of an atom of sodium?

5) How many electrons are there in the outer shell of an atom of chlorine?

6) How many electrons will the N shell hold?

7) Which electron shell holds the least number of electrons?

8) Which electron shells hold the greatest number of electrons?

9) Use a dictionary to define the word *stable*.

10) When an atom has a "+" charge, what do you know about the number of protons compared to the number of electrons?

5.3 COMPOUNDS HAVE FORMULAS

Many compounds have just two atoms. Table salt is one example we have already seen. Other compounds have different atoms and different numbers of atoms. Scientists use the symbols for the elements to write a *FORMULA* for each compound. A chemical formula tells what elements and how many atoms of each element are in a compound. In many ways, a formula is like a recipe. A recipe lists the ingredients and how much of each is needed to make something.

A Recipe
1 egg
1 cup flour
1 teaspoon salt
2 teaspoons sugar
1 cup water

A Chemical Formula
NaCl

A simple formula like NaCl means that there is one atom of sodium (Na) and one atom of chlorine (Cl). One unit of table salt, then, contains a total of two atoms.

Another example is the compound, water. Water has the formula H_2O. This formula tells us that there are two atoms of hydrogen and one atom of oxygen. A molecule of water, then, contains a total of three atoms $(2 + 1 = 3)$. The number that tells how many atoms of an element there are is called a *SUBSCRIPT*. It is written after the element to which it applies. In addition to being written after the element, it is generally written below the symbol.

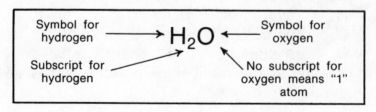

▶ Study the following examples:

1) What elements and how many atoms of each element are in the compound $BaCl_2$?

Symbol	Element	Subscript	Number of Atoms
Ba	barium	none	1
Cl	chlorine	2	+ 2

3 Total atoms

2) What elements and how many atoms of each element are in the compound CH_4?

Symbol	Element	Subscript	Number of Atoms
C	carbon	none	1
H	hydrogen	4	+ 4

5 Total atoms

3) What elements and how many atoms of each element are in the compound HNO_3?

Symbol	Element	Subscript	Number of Atoms
H	hydrogen	none	1
N	nitrogen	none	1
O	oxygen	3	+ 3

5 Total atoms

4) What elements and how many atoms of each element are in the compound $NaHCO_3$?

Symbol	Element	Subscript	Number of Atoms
Na	sodium	none	1
H	hydrogen	none	1
C	carbon	none	1
O	oxygen	3	+ 3

6 Total atoms

5) What elements and how many atoms of each element are in the compound $C_{12}H_{22}O_{11}$?

Symbol	Element	Subscript	Number of Atoms
C	carbon	12	12
H	hydrogen	22	22
O	oxygen	11	+ 11
			45 Total atoms

▶ *Practice.*

1) What information does a formula tell you about a compound?

2) How is a formula like a recipe?

3) What is a subscript?

4) Write a formula for a compound containing one atom of potassium and one atom of chlorine.

5) Write a formula for a compound containing one atom of magnesium and two atoms of iodine.

6) For each of the following compounds, tell what elements are in the compound, what the subscript is for each element, how many atoms of each element there are, and the total number of atoms in the compound.

a) $AlBr_3$ b) C_2H_2

c) $ZnCl_2$ d) BaO

e) HCl f) $K_2Cr_2O_7$

g) $KClO_3$ h) $AlCl_3$

i) $MgCl_2$ j) CO_2

k) H_2SO_4 l) CaO

5.4 COMPOUNDS CONTAINING RADICALS

The formulas for some compounds contain groups of two or more atoms which *act* as if they were one element. These groups of atoms are called *RADICALS*. One common substance whose formula contains a radical is household lye, a chemical used to clean drains. The formula for lye is NaOH. The OH is an example of a radical. It contains one atom of oxygen and one atom of hydrogen. The chemical name for this radical is the "hydroxyl" radical. Other examples of radicals and their names are listed in the following chart.

Radical	Name
SO_4	sulfate
ClO_3	chlorate
NO_3	nitrate
CO_3	carbonate

Compounds which contain more than one radical are written with the radical in parentheses, with a subscript outside of the parentheses. Look at these examples:

$$Ba(OH)_2 \text{ means } Ba \begin{smallmatrix} (OH) \\ \\ (OH) \end{smallmatrix} \qquad Al(OH)_3 \text{ means } Al \begin{smallmatrix} (OH) \\ -(OH) \\ (OH) \end{smallmatrix}$$

When formulas contain radicals with subscripts, the subscripts MULTIPLY the number of atoms inside the parentheses. Study the examples on the next page.

1) What elements and how many atoms of each are in the compound $Al(OH)_3$?

Symbol	Element	Subscript	Radical Subscript	Number of Atoms
Al	aluminum	none	not in a radical	1
O	oxygen	none	3	3 (3 × 1)
H	hydrogen	none	3	+ 3 (3 × 1)
				7 Total atoms

2) What elements and how many atoms of each are in the compound $Ba(NO_3)_2$?

Symbol	Element	Subscript	Radical Subscript	Number of Atoms
Ba	barium	none	not in a radical	1
N	nitrogen	none	2	2 (2 × 1)
O	oxygen	3	2	+ 6 (2 × 3)
				9 Total atoms

3) What elements and how many atoms of each are in the compound $Al(NO_3)_3$?

Symbol	Element	Subscript	Radical Subscript	Number of Atoms
Al	aluminum	none	not in a radical	1
N	nitrogen	none	3	3 (3 × 1)
O	oxygen	3	3	+ 9 (3 × 3)
				13 Total atoms

4) What elements and how many atoms of each are in the compound $Al_2(CO_3)_3$?

Symbol	Element	Subscript	Radical Subscript	Number of Atoms
Al	aluminum	2	not in a radical	2
C	carbon	none	3	3 (3×1)
O	oxygen	3	3	+ 9 (3×3)
				14 Total atoms

▶ *Practice.*

For each of the following compounds, list the elements, the number of atoms of each element, and the total number of atoms in the compound.

1) $Ca(OH)_2$ 2) $Ba(OH)_2$

3) $Mg(NO_3)_2$ 4) $Ca(NO_3)_2$

5) $Al_2(SO_4)_3$ 6) $BaSO_4$

5.5 NAMING COMPOUNDS

Just as each element has a symbol and a name, every compound has a formula and a chemical name. However, the name for a compound is actually two names written together. The name of a compound is similar to a person's name — there is a first name and a last name.

Although there is no easy set of rules which will allow you to name all compounds correctly, there are a few rules by which you can name most of the common compounds that you will see.

5.6 NAMING COMPOUNDS CONTAINING ONLY TWO ELEMENTS

To apply the rules, you must first find out how many elements are in the compound. This does NOT mean how many *atoms*, only how many different *elements* there are in the compound. If there are two elements only, use these rules for naming the compound.

1) The first name of the compound is the same as the name of the first element in the formula.

2) The second name is the name of the second element in the formula with the ending changed to "ide."

   For this second rule, use these names:

For:	*Use:*
chlorine	chlor**ide**
iodine	iod**ide**
fluorine	fluor**ide**
bromine	brom**ide**
oxygen	ox**ide**
sulfur	sulf**ide**

Study the following examples of naming compounds containing only two elements.

1) What is the name of the compound whose formula is NaCl?

   First name is the name of the first element: sodium

   Second name is *chlorine* changed to: chloride

   Compound name is: **sodium chloride**.

   Sodium chloride is the chemical name for salt.

2) What is the name of the compound whose formula is LiF?

First name is the name of the first element: lithium

Second name is *fluorine* changed to: fluoride

Compound name is: **lithium fluoride**.

3) What is the name of the compound whose formula is BaO?

First name is the name of the first element: barium

Second name is *oxygen* changed to: oxide

Compound name is: **barium oxide**.

4) What is the name of the compound whose formula is Ag_2S?

First name is the name of the first element: silver (The subscript has no effect on the name.)

Second name is *sulfur* changed to: sulfide

Compound name is: **silver sulfide**.

5) What is the name of the compound whose formula is Al_2S_3?

First name is the name of the first element: aluminum (The subscript has no effect on the name.)

Second name is *sulfur* changed to: sulfide (The subscript has no effect on the name.)

Compound name is: **aluminum sulfide**.

▶ **Practice.**

Name the following compounds which contain only two elements.

1) $CaBr_2$ 2) AgI

3) $AlCl_3$ 4) MgS

5) HCl 6) $NaBr$

7) $CaCl_2$ 8) BaI_2

9) CaF_2 10) MgO

5.7 NAMING COMPOUNDS WITH MORE THAN TWO ELEMENTS

Most compounds containing more than two elements have a radical in their formula. Any formula containing one of the following listed radicals can be named using the next rule.

1) The first name of the compound is the name of the first element in the formula.

2) The second name is one of the following:

Radical	Name
PO_4	phosphate
CO_3	carbonate
SO_4	sulfate
NO_3	nitrate
OH	hydroxide

Remember that in some formulas the radical will be enclosed in parentheses and have a subscript. In other formulas there will be no subscript. In either case, the subscripts outside of the parentheses have no effect on the name of the compound. Be sure, however, that the subscript in the radical matches those in the chart just given.

For example, CO_3 is the radical carbonate, while CO_2 is not. SO_4 is the sulfate radical, but SO_3 is not.

Study the following examples of naming compounds with more than two elements.

1) What is the name of the compound whose formula is $CaCO_3$?

First name is the name of the first element: calcium

Second name comes from the CO_3 radical: carbonate

Compound name is: **calcium carbonate**.

2) What is the name of the compound whose formula is $Al(OH)_3$?

First name is the name of the first element: aluminum

Second name comes from the OH radical: hydroxide

Compound name is: **aluminum hydroxide**.

3) What is the name of the compound whose formula is Na_2SO_4?

First name is the name of the first element: sodium

Second name comes from the SO_4 radical: sulfate

Compound name is: **sodium sulfate**.

4) What is the name of the compound whose formula is $AlPO_4$?

First name is the name of the first element: aluminum

Second name comes from the PO_4 radical: phosphate

Compound name is: **aluminum phosphate**.

▶ **Practice.** Name the following compounds.

1) $Ba(OH)_2$ 2) $MgSO_4$ 3) $Ca(OH)_2$

4) $Al_2(SO_4)_3$ 5) $BaSO_4$ 6) KNO_3

7) K_2CO_3 8) Na_2CO_3 9) $Ca_3(PO_4)_2$

10) $Al_2(CO_3)_3$ 11) Ag_2SO_4 12) $ZnSO_4$

13) KOH   14) $Mg(OH)_2$ 15) $Al(NO_3)_3$

5.8 COMPOUNDS CALLED ACIDS

Does the thought of licking a lemon make your mouth water? Lemons contain an *ACID* known as citric acid. Acids are a special group of compounds which have a sour taste. Scientists have found that all acids contain the element hydrogen (H). Some of the more common acids are listed in the following table.

Common Acids	
Acid Name	**Where Found**
Citric acid	Citrus fruits
Acetic acid	Vinegar
Lactic acid	Sour milk
Boric acid	Eye washes
Acetylsalicylic acid	Aspirin tablets
Tannic acid	Tea

5.9 COMPOUNDS CALLED BASES

Another special group of compounds are called *BASES*. Bases have a bitter taste. They generally have a slippery feel when touched. Have you ever accidentally gotten soap in your mouth while washing? If you have, you know how bitter it tastes. Scientists have found that all bases contain the radical OH, the hydroxyl radical.

Common Bases	
Base Name	**Where Found**
Sodium hydroxide	Household lye
Magnesium hydroxide	Milk of magnesia
Ammonium hydroxide	Household ammonia

Naming the Acids and Bases

Study the chart of common bases above. Most of the names for *bases* follow the rules that you have learned in this chapter. However, most *acids* cannot be named by the rules. Here is a list of the chemical names and formulas for acids used in the laboratory:

Acids and Their Names	
HCl	Hydrochloric acid
H_2SO_4	Sulfuric acid
HNO_3	Nitric acid
H_3BO_3	Boric acid
$HC_2H_3O_2$	Acetic acid

SUMMARY OF CHAPTER 5

1) A compound is a chemical combination of two or more elements.

2) Electrons are located in shells around the nucleus of an atom.

3) Atoms tend to share, borrow, or lend electrons.

4) Compounds are formed when atoms share, borrow, or lend their electrons to other atoms.

5) Atoms are more stable when their shells contain the proper number of electrons.

6) Atoms attract each other in compounds because of opposite charges.

7) A formula lists the elements in a compound and the number of atoms for each element.

8) Subscripts show the number of atoms of an element that are present in a compound.

9) When there is no subscript beside an element's symbol in a formula, 1 atom of that element is in the compound.

10) A radical is a group of atoms that act like one element.

11) Compounds have two names.

12) Acids are compounds which contain hydrogen. They have a sour taste.

13) Bases are compounds which contain the hydroxyl radical. They have a bitter taste and a slippery feel.

• CHAPTER REVIEW EXERCISES •

Vocabulary Words

compound   radical   electron shell
subscript   formula   ion
acid   base

▶ Copy the sentences. Fill in the missing words.

1) A chemical combination of two or more elements is called a ____.

2) OH is an example of a ____.

3) The chemical name for $BaCl_2$ is ____ ____.

4) The total number of atoms in $AlCl_3$ is ____.

5) Acids have a ____ taste.

▶ Write the letter of the best answer for each item.

6) In the formula H_2CO_3, the number of atoms of carbon is:
   a) 2   b) 1   c) 3

7) The compound sodium bromide has the formula:
   a) SoBr   b) NaBr   c) NaB

8) The name of the compound MgS is:
   a) manganese sulfide.
   b) magnesium sulfide.
   c) magnesium sulfur.

9) How many atoms of hydrogen are in the compound $NaHCO_3$?
   a) 1   b) none   c) 3

10) Which of the following is the formula for salt?
    a) NaCl   b) $BaCl_2$ c) NaI

Computer Program

▶ The following program will print a chart listing some common household chemicals which you have studied. The chart also lists their chemical names. When you run the program, study the chart first. Then after you press the Return key, the computer will give you a practice test.

```
5 LET C=0
10 CLS
20 PRINT "STUDY THE FOLLOWING CHART"
30 PRINT:PRINT
40 READ N$,C$
50 IF N$="LAST" THEN 100
60 PRINT N$;TAB(18);C$
70 PRINT
80 READ N$,C$
90 GOTO 50
100 INPUT "PRESS RETURN FOR TEST";Z$
110 CLS
120 RESTORE
130 READ N$,C$
140 IF N$="LAST" THEN 250
150 PRINT "WHAT IS THE CHEMICAL NAME FOR: "
160 PRINT
170 PRINT N$,
180 INPUT R$
190 IF R$=C$ THEN PRINT "CORRECT":C=C+1:GOTO 210
200 PRINT "INCORRECT"
210 INPUT "PRESS RETURN TO CONTINUE ";Z$
220 CLS
230 READ N$,C$
240 GOTO 140
250 PRINT:PRINT
260 PRINT "TEST COMPLETED"
270 PRINT
280 PRINT "YOU GOT ";C;" CORRECT."
290 END
300 DATA EPSOM SALT, MAGNESIUM SULFATE
310 DATA TABLE SALT, SODIUM CHLORIDE
320 DATA RUST, IRON OXIDE
330 DATA BAKING SODA, SODIUM BICARBONATE
340 DATA WASHING SODA, SODIUM CARBONATE
350 DATA LAST, LAST
```

Chapter 6

How Matter Changes

Chapter Goals:

1. To list the four kinds of reactions.

2. To describe three devices for heating chemicals.

3. To state one method of causing reactions.

4. To state the law of conservation of matter.

5. To translate a chemical equation into words.

6. To explain why chemical equations must be balanced.

7. To describe what is meant by a solution.

8. To explain a method of separating a mixture by dissolving one of the substances.

6.1 WHAT IS A REACTION?

During the Middle Ages, people known as *ALCHEMISTS* tried to change different materials into gold or other precious metals. Imagine being able to change dirt or sand into solid gold!

Unfortunately for the alchemists, they never succeeded. In fact, no one to this day has been able to change any substance into gold. Although we still can't change one element into another, we *can* take elements and form different compounds. Whenever a substance is changed into a different substance, we say that it has undergone a *REACTION*.

6.2 HOW DO REACTIONS OCCUR?

Substances don't always react. In fact, many elements and compounds can be mixed together and nothing happens at all. For some reactions, it is necessary to add special chemicals. For other reactions the substances must be heated. Heating is the most common method of causing reactions. In the laboratory, three types of heat sources can be used.

A Hot Plate Gives Heat.

The first heat source is an electric *HOT PLATE*. It usually has a flat surface on which beakers can be placed to be heated. A heat control is located on the front of the hot plate to control the temperature.

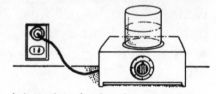

A hot plate for heating materials

An Alcohol Burner Gives Heat.

Another method of heating is to use the *ALCOHOL BURNER*. An alcohol burner has a base filled with alcohol and a wick usually made from a cotton material. Alcohol burners are often supplied in home chemistry kits. Never use an alcohol burner without proper adult supervision. There is a danger of spilling the alcohol and causing a fire.

An alcohol burner

A Bunsen Burner Gives Heat.

The third type of heat source is the most common type found in the laboratory. It is called a *BUNSEN BURNER*. A Bunsen burner burns natural gas (the same gas that is burned on kitchen stoves at home). Bunsen burners are efficient at heating test tubes, but they can also be dangerous because the flame is often difficult to see.

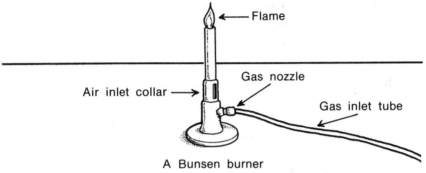

A Bunsen burner

Be Careful!

When you observe chemical reactions, always wear protective goggles to prevent chemicals from getting into your eyes. Also, when you heat chemicals, always point the open end of the test tube away from any other person.

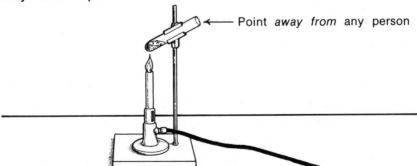

The most important rule when working with chemicals is to BE CAREFUL! Even so-called harmless chemicals can be dangerous at times. Experiments should be done only under adult supervision.

6.3 TYPES OF REACTIONS

All of the possible reactions of matter can be grouped into four major categories or types. First, a *SYNTHESIS* reaction is one where two or more elements combine to form one compound. Second is the *ANALYSIS* reaction. In this type of reaction, one compound separates into two or more elements. Third is the *SIMPLE REPLACEMENT* reaction, where one element is replaced in a compound by another. Lastly, there is the *DOUBLE REPLACEMENT* reaction, where all the elements in two compounds are exchanged.

Reactions	
Type	**Form**
1. Synthesis	A plus B makes AB
2. Analysis	AB makes A plus B
3. Simple Replacement . . .	A plus BC makes B plus AC
4. Double Replacement . . .	AB plus CD makes AD plus CB

6.4 REPRESENTING REACTIONS

Scientists have designed a method of using symbols to describe reactions, rather than words. The method uses *CHEMICAL EQUATIONS* to describe reactions. A chemical equation is a statement describing a reaction which takes place. The chemicals involved are described by using their symbols and formulas. Here is an example of a simple reaction described by using a chemical equation. Below the equation you can see the same description in words.

$$HCl \quad + \quad NaOH \quad \longrightarrow \quad NaCl \quad + \quad H_2O$$

hydrogen plus sodium makes sodium plus water
chloride   hydroxide   chloride

The chemical or chemicals on the left side of the arrow are called *REACTANTS*. They are the chemicals which are going to react together. The chemicals on the right side of the arrow are called the *PRODUCTS*. They are the chemicals that form after the reaction. The four types of reactions can now be represented using letters to show the position of different elements.

Reactions	
Type	**Form**
1. Synthesis.............	$A + B \rightarrow AB$
2. Analysis..............	$AB \rightarrow A + B$
3. Simple Replacement ...	$A + BC \rightarrow B + AC$
4. Double Replacement ...	$AB + CD \rightarrow AD + CB$

▶ *Practice.* Write the answers.

1) What did alchemists try to do?
2) Name three devices for heating chemicals in the laboratory.
3) List the four types of reactions.
4) For each chemical equation below, write the reactants. Then list the products for each reaction.
   a) $Fe + S \rightarrow FeS$
   b) $H_2SO_4 + Zn \rightarrow ZnSO_4 + H_2$
   c) $MgS \rightarrow Mg + S$
   d) $AgNO_3 + NaCl \rightarrow NaNO_3 + AgCl$
5) Write a description for the reaction below. Use the symbols for chemical equations.
   Calcium plus oxygen makes calcium oxide.
6) Write the following chemical equation in words.
   $$Mg + S \rightarrow MgS$$

6.5 THE LAW OF CONSERVATION OF MATTER

In any normal chemical reaction, the reactants present *before* a reaction can be quite different from the products present *after* the reaction. However, the same atoms that are present before the reaction are also present after the reaction. Different compounds may be formed, but the same atoms are always there. Also, in any normal reaction, the mass of the reactants is equal to the mass of the products. Another way of saying this is summarized in the LAW OF CONSERVATION OF MATTER:

Law of Conservation of Matter

In any ordinary reaction, matter is neither created nor destroyed.

6.6 BALANCING EQUATIONS

Because of the law of conservation of matter, the same number and kinds of atoms must be present in the reactants and the products. In addition, the formula for a compound cannot be changed. In order to satisfy the law of conservation of matter, a chemical equation must be *BALANCED* to keep the number of atoms the same. Look at the following reaction. It shows that hydrogen plus oxygen makes water.

$$H_2 + O_2 \rightarrow H_2O$$

To write a balanced equation, start with the correct formulas for each of the reactants and products. Notice that the formulas for hydrogen and oxygen each have 2 atoms. Hydrogen and oxygen are examples of certain gases which normally form molecules containing two atoms of the element. The formulas for all three substances in the equation are correct as written.

Now, count the number of atoms of each element ON EACH SIDE OF THE ARROW.

$$H_2 + O_2 \longrightarrow H_2O$$

H — 2 atoms	H — 2 atoms
O — 2 atoms	O — 1 atom
Total of 4 atoms	Total of 3 atoms

Something is wrong because one atom of oxygen is missing from the right side of the equation.

The law of conservation of matter says that atoms do not disappear in reactions. To make the two sides come out even, we can BALANCE the equation. We can place numbers called COEFFICIENTS before the molecules. Coefficients multiply the numbers of atoms in the formula. By changing these coefficients, equations can be balanced. Here is the original equation for forming water:

$$H_2 + O_2 \longrightarrow H_2O$$

By placing coefficients properly, it can be balanced like this:

$$2H_2 + O_2 \longrightarrow 2H_2O$$

H — 4 atoms	H — 4 atoms
O — 2 atoms	O — 2 atoms
Total of 6 atoms	Total of 6 atoms

The equation is now balanced. The coefficients tell how many molecules of each substance are present in a reaction. If no coefficient is present, as in the oxygen above, the coefficient is read as a 1. The above reaction can be described in words as:

2 molecules of hydrogen plus 1 molecule of oxygen make 2 molecules of water.

▶ **Practice.** Study the equation. Then answer the questions.

$$2Na + Cl_2 \rightarrow 2NaCl$$

1) What are the reactants?
2) What is the product?
3) How many atoms of chlorine are on the left side of the arrow?
4) How many atoms of chlorine are on the right side of the arrow?
5) How many atoms of sodium are on the left side of the arrow?
6) How many atoms of sodium are on the right side of the arrow?

6.7 A SYNTHESIS REACTION

A simple example of a synthesis reaction involves iron and sulfur, two common elements. Iron, whose symbol is Fe, is a metal used in the making of steel. Sulfur, whose symbol is S, is a yellow powder that is sometimes used in medicines.

Iron   —   Fe   —   a silver gray metal

Sulfur   —   S   —   a yellow powder

If iron in the form of iron filings (tiny slivers of iron) is mixed with sulfur powder, no reaction takes place. The combination of iron and sulfur is now called a *MIXTURE*. A mixture is formed when substances are simply stirred together and no new compound is formed. No reaction takes place when a mixture is formed.

A mixture of iron and sulfur

Since no reaction takes place, a mixture still keeps the individual properties of the separate substances. In fact, we can separate the iron and sulfur by placing a magnet in the mixture. Because the iron is attracted to a magnet, but not sulfur, the mixture can be separated quite easily.

Iron filings are attracted to the magnet.

If the mixture of iron and sulfur is heated using a Bunsen burner, a reaction will occur. If the correct amount of each substance is present, a new compound called *iron sulfide* will be formed. Here is the chemical equation for this synthesis reaction:

$$Fe \quad + \quad S \quad \longrightarrow \quad FeS$$

iron plus sulfur makes iron sulfide

One molecule of iron plus one molecule of sulfur makes one molecule of iron sulfide.

Iron sulfide has different properties from either iron or sulfur. In fact, the properties are completely different. If a magnet is placed near the iron sulfide, it will not be attracted to it. The characteristic yellow color of sulfur is also gone. The color of iron sulfide is gray-black. The properties have changed because a new compound has been formed.

6.8 AN ANALYSIS REACTION

An interesting analysis reaction can be produced when mercuric oxide, an orange-colored, poisonous compound, is heated in a test tube. This orange powder forms a silver-gray liquid and a gas which

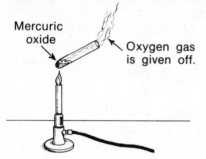

escapes from the test tube when heated. The liquid is mercury (Hg), the element which is used in thermometers. The gas that is given off is oxygen, the gas necessary for life.

The analysis reaction is written in chemical equation form as:

$$2HgO \rightarrow 2Hg + O_2\uparrow (g)$$

2 molecules of mercuric oxide make 2 atoms of mercury and 1 molecule of oxygen.

The letter (g) after the O_2 and the upward pointing arrow indicate that a gas is formed. This gas is given off by the reaction.

6.9 A SIMPLE REPLACEMENT REACTION

Hydrogen gas is often made in the laboratory. Hydrochloric acid is used, along with zinc, a metal used in the making of batteries. When the zinc metal is placed into the hydrochloric acid, a reaction immediately takes place, causing bubbles to form. The bubbles that form are hydrogen gas. Here is the chemical equation for this simple replacement reaction.

$$Zn + 2HCl \rightarrow ZnCl_2 + H_2\uparrow (g)$$

One atom of zinc plus 2 molecules of hydrochloric acid make 1 molecule of zinc chloride and 1 molecule of hydrogen gas.

6.10 A DOUBLE REPLACEMENT REACTION

An example of a double replacement reaction can be shown by the reaction between table salt, NaCl, and a substance known as silver nitrate, $AgNO_3$. The reaction is shown in the following chemical equation.

$$NaCl + AgNO_3 \longrightarrow NaNO_3 + AgCl\downarrow (s)$$

Sodium chloride plus silver nitrate make sodium nitrate plus silver chloride.

The letter (s) after the AgCl and the downward pointing arrow mean that this substance is a solid which settles to the bottom of the test tube. When a solid like this is formed, it is called a *PRECIPITATE*.

6.11 SOLUTIONS

Many reactions involve compounds that have been *DISSOLVED* in other liquids. An example of dissolving is when salt is placed in water. The salt seems to disappear into the water. The salt is still there, which can be shown by tasting the water. The large crystals of salt have been dissolved into smaller particles that become invisible in the water. When a substance is dissolved in a liquid, the result is called a *SOLUTION*. The substance that dissolves is called the *SOLUTE*. The substance in which the dissolving is done is called the *SOLVENT*.

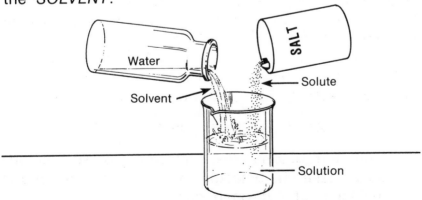

INVESTIGATION 5

Purpose: To separate a mixture by using the principle of dissolving.

A. Materials

Quantity:	*Item Description:*
2 grams	Sand
2 grams	Salt
2	Beakers or jars
1	Piece of filter paper or paper towel
200 mL	Water
1	Stirrer

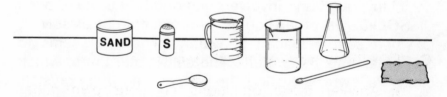

B. Procedure

1) On a sheet of paper, mix the salt with the sand. Mix thoroughly.

2) Describe the resulting mixture.

3) Place the mixture in the beaker and cover with tap water.

4) Stir the solution to help the salt dissolve.

5) Place the filter paper across the top of a second beaker.

6) Slowly pour the solution into the second beaker through the filter paper.

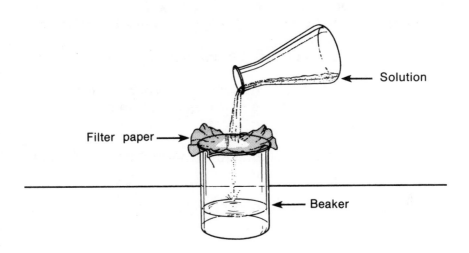

C. Questions and Interpretation

▶ Answer these questions on your own paper.

1) What happens to the salt when the water is added to the mixture?

2) Does the sand dissolve? How do you know?

3) Which material remains on the filter paper?

4) Where is the salt after you pour the solution into the second beaker? How do you know?

5) Take the beaker and let it sit for several days. What do you find on the bottom of the beaker after the water has evaporated?

6) What is the solvent in this investigation?

7) What is the solute in this investigation?

SUMMARY OF CHAPTER 6

1) A reaction occurs when a substance is changed into a new substance.

2) Many reactions can be caused by heating.

3) A Bunsen burner, an alcohol burner, and a hot plate are devices used to heat chemicals in the laboratory.

4) Always perform chemical experiments with adult supervision.

5) When working with chemicals, always use protective goggles.

6) When heating chemicals in a test tube, be sure to point the mouth of the test tube AWAY from any people.

7) Chemical reactions are described using chemical equations.

8) *Reactants* are substances which are going to react. They are written on the left side of the arrow in a chemical equation.

9) The *products* are substances formed after a reaction has taken place. They are written on the right side of the arrow in a chemical equation.

10) The law of conservation of matter says that matter is neither created nor destroyed in normal reactions.

11) A chemical equation must be balanced by placing coefficients in front of the formulas of a chemical equation.

12) A *mixture* does not form new substances.

13) Mixtures can be separated into the original substances.

14) A *solution* is formed when a substance is dissolved in a liquid.

15) The substance which dissolves is called the *solute*.

16) The substance in which something dissolves is called the *solvent*.

Vocabulary Words

alchemist
reaction
hot plate
alcohol burner
Bunsen burner
synthesis
chemical equation
reactant
coefficient
product
balanced equation
solution
precipitate
dissolve
solvent
solute
analysis reaction
simple replacement reaction
double replacement reaction
law of conservation of matter

• CHAPTER REVIEW EXERCISES •

▶ Copy the sentences. Fill in the missing words.

1) People who tried to change different substances into gold are called ____.

2) A heating device which uses natural gas is called a ____ ____.

3) A reaction where two or more elements combine to form one compound is called a ____ reaction.

4) The substance which dissolves is called the ____.

5) The substances on the left of the arrow in a chemical equation are called the ____.

▶ Write the letter of the best ending for each sentence.

6) A solid which forms and settles to the bottom of a test tube during a reaction is a
a) precipitate.   b) solvent.   c) solution.

7) The numbers placed in front of substances in a chemical equation are called
a) formulas.   b) subscripts.   c) coefficients.

8) In the equation: $2H_2 + O_2 \rightarrow 2H_2O$ O_2 is a
a) reactant.   b) product.   c) precipitate.

9) When dry substances are simply stirred together, they are called a
a) precipitate.   b) solution.   c) mixture.

10) In the equation: $2H_2O \rightarrow 2H_2 + O_2 \uparrow (g)$ the (g) and the upward arrow mean that oxygen is a
a) gas.   b) liquid.   c) precipitate.

Computer Program

▶ The following computer program will give you a test on the different types of chemical reactions.

```
 5 C=0
10 CLS
20 READ S$,T$
30 IF S$="LAST" THEN 240
40 PRINT "WHAT TYPE OF REACTION IS THIS:"
50 PRINT:PRINT
60 PRINT S$
65 PRINT:PRINT "TYPE 'S' FOR SYNTHESIS"
75 PRINT
80 PRINT "TYPE 'A' FOR ANALYSIS"
85 PRINT
90 PRINT "TYPE 'SR' FOR SIMPLE REPLACEMENT"
95 PRINT
100 PRINT "TYPE 'DR' FOR DOUBLE REPLACEMENT"
105 PRINT
110 PRINT
120 INPUT "YOUR ANSWER: ";R$
130 IF R$=T$ THEN PRINT "CORRECT":C=C+1:GOTO 150
140 PRINT "INCORRECT"
150 PRINT:PRINT
160 PRINT "PRESS RETURN TO CONTINUE"
170 INPUT R1$
180 GOTO 10
190 DATA "A + B --> AB","S"
200 DATA "AB --> A + B","A"
210 DATA "A + BC --> AC + B","SR"
230 DATA "AB + CD --> AD + CB","DR"
240 DATA "LAST", "LAST"
250 CLS:PRINT "NUMBER CORRECT OUT OF 4 --> ";C
260 END
```

Chapter 7

Force and Motion

Chapter Goals:

1. To be able to calculate the speed of an object.

2. To describe what is meant by *average speed*.

3. To solve the distance, rate, and time formula for each of the variables.

4. To obtain information from a graph of distance and time.

5. To use a graph to predict distance.

6. To define *force*, *acceleration*, *deceleration*, *friction*, and *gravity*.

7. To describe Newton's Three Laws of Motion.

8. To state the Universal Law of Gravitation.

7.1 MOTION

Let's imagine you have just taken a trip from the city of Startsville to the city of Endsville. Your flight began at 8:00 P.M. You arrived in Endsville at 11:00 P.M. The distance between those two cities is 2100 miles. How long did it take for this trip?

7.2 TIME

To calculate the duration of the flight in hours, subtract the departure time of 8 from the arrival time of 11 to obtain:

11:00 Arrival time

−8:00 Departure time

3:00 Hours travel time

Departure   Arrival
Time   Time

7.3 SPEED

We can now ask, "How fast was the airplane traveling? What was its speed?" *SPEED* is the distance traveled per unit of time. This can be expressed in mathematical terms in the following way:

Speed = Distance ÷ Time

Here is how we calculate the speed for this example:

Speed = 2100 miles ÷ 3 hours
Speed = 700 miles / hour

The result of this calculation is read as 700 miles per hour. This means that during each hour the plane traveled an average of 700 miles.

Speed does not have to be measured in miles per hour. For example, consider a foot race at a track meet where the distance around the track is 100 meters. A runner completes the race in 15 seconds. What was the runner's speed?

Speed = Distance ÷ Time
Speed = 100 meters ÷ 15 seconds
Speed = 6.67 meters /second

This answer is read 6.67 meters per second. Notice that the answer has been rounded to the nearest hundredths place. In other words, the runner covers an average distance of 6.67 meters in 1 second.

Average Speed

In the examples of the airplane and the runner, the speeds are actually *average* speeds. It is unlikely that the airplane maintained a constant speed of 700 miles per hour during the entire flight. Although the *average* speed is 700 miles per hour, the *actual* speed at any particular moment could be more or less than this speed.

▶ *Practice.* Copy the following chart. Calculate the average speed for each of these examples. The first one is completed for you.

Distance Traveled		Time	Average Speed
1)	30 miles	5 hours	6 miles/hour
2)	100 yards	13 seconds	
3)	10 cm	5 seconds	
4)	380 km	2 hours	
5)	3825 feet	30 minutes	
6)	15 inches	4 hours	
7)	82 miles	10 hours	
8)	10,000 meters	36 minutes	
9)	23 feet	6 minutes	
10)	120 km	2 hours	
11)	600 feet	10 minutes	
12)	40 miles	2 hours	

7.4 CALCULATING DISTANCE

The formula for finding speed can be used to calculate the distance when the average speed and the time are known. The formula:

$$\text{SPEED} = \text{DISTANCE} \div \text{TIME}$$

can be rearranged to solve for the distance in the following way:

$$\text{DISTANCE} = \text{SPEED} \times \text{TIME}$$

Imagine that a jogger runs for 3 hours at an average speed of 6 miles per hour. The distance traveled can be found by using the formula:

$$\text{DISTANCE} = \text{SPEED} \times \text{TIME}$$

$$\text{DISTANCE} = \frac{6 \text{ miles}}{1 \text{ hour}} \times \frac{3 \text{ hours}}{1}$$

$$\text{DISTANCE} = 18 \text{ miles}$$

The same formula can also be solved for the time.

$$\text{TIME} = \text{DISTANCE} \div \text{SPEED}$$

Example 1: A white line is to be painted down the middle of a road. The distance to be painted is 60 meters long. If the painter paints at a speed of 5 meters per minute, how much time will it take to complete the line?

Solution:
$$\text{TIME} = \text{DISTANCE} \div \text{SPEED}$$

$$\text{TIME} = 60 \text{ meters} \div \frac{5 \text{ meters}}{1 \text{ minute}}$$

$$\text{TIME} = \frac{60 \text{ meters}}{1} \times \frac{1 \text{ minute}}{5 \text{ meters}}$$

$$\text{TIME} = \frac{60 \text{ minutes}}{5}$$

$$\text{TIME} = 12 \text{ minutes}$$

Example 2: A man walks along the beach at a speed of 50 meters per minute. How much time does it take for him to walk a distance of 210 meters?

Solution: TIME = DISTANCE ÷ SPEED

$$TIME = 210 \text{ meters} \div \frac{50 \text{ meters}}{1 \text{ minute}}$$

$$TIME = \frac{210 \text{ meters}}{1} \times \frac{1 \text{ minute}}{50 \text{ meters}}$$

$$TIME = \frac{210 \text{ minutes}}{50}$$

$$TIME = 4.2 \text{ minutes}$$

If necessary, round your answers to the nearest tenths place.

▶ **Practice.** Find the answer for each word problem. Use one of these formulas: (Distance = Speed × Time) or (Time = Distance ÷ Speed).

1) Calculate the distance a train travels if its speed is 60 kilometers per hour and the time is 2 hours.

2) A student rides his ten-speed bike to school. If his school is 5 miles from his home, and he travels at a rate of 16 miles per hour, how much time is needed for this trip?

3) If it takes a plane 5 hours at a rate of speed of 850 miles per hour to travel from point A to point B, what is the distance between these two points?

4) A rocket can travel at a rate of 18,000 miles per hour. How far will the rocket travel in 4.5 hours?

5) A man rode on a motorcycle for two hours. If his speed was 45 miles per hour, how far did he travel?

7.5 DESCRIBING MOTION USING A GRAPH

Information about speed, distance, and time can be shown in a data table. Suppose that three different automobiles traveled for 5 hours. The following data table shows the elapsed time and the distance traveled by the three autos.

Elapsed Time in Hours	Distance Traveled in km		
	Auto 1	Auto 2	Auto 3
1	50	40	25
2	100	80	50
3	150	120	75
4	200	160	100
5	250	200	125

The same information that is in the data table above can also be shown on a line graph.

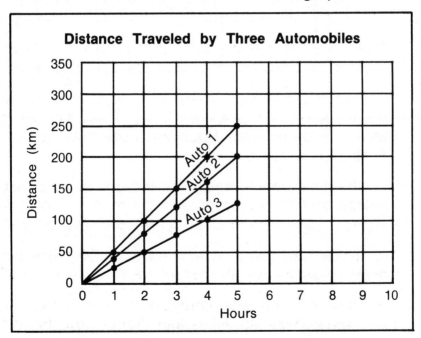

7.6 USING GRAPHS TO OBTAIN INFORMATION

One thing which we can now do with the graph is to determine the speed at times that are not listed in the data table. For example, suppose we want to know the distance traveled by Auto 3 after 4.5 hours. Draw a vertical line at 4.5 hours up to the line for Auto 3. At this point draw a horizontal line to the distance axis and read the distance traveled. This method is shown below.

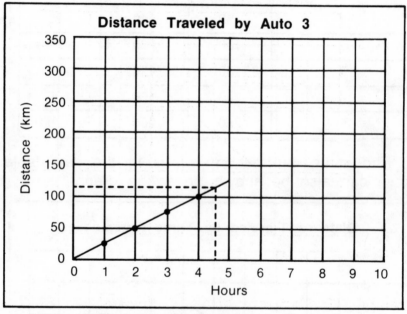

Steps for determining distance traveled by Auto 3 at a time of 4.5 hours:

1. Find the time of 4.5 hours. This is half-way between 4 hours and 5 hours.

2. Draw a vertical line up to the line for Auto 3.

3. At the intersection, draw a horizontal line to the distance axis.

4. Read the distance on the scale.

5. The distance is approximately 112 kilometers.

Making Predictions

A graph can also be used to predict where the auto would be at some future time which is not shown in the original data table. This assumes, however, that the autos continued at a constant rate of speed.

In the following graph, the line for Auto 2 has been extended by using a dotted line.

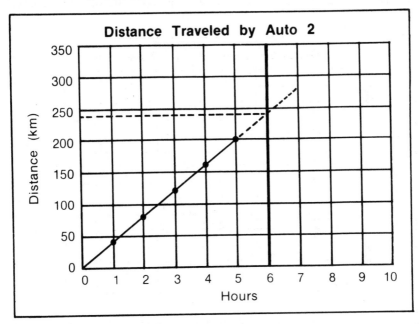

Distance Traveled by Auto 2

Example: Predict the distance traveled by Auto 2 at a time of 6 hours.

Steps: 1. Draw a vertical line from the 6-hour line to the top of the graph.

2. Extend the graph line through the vertical line.

3. From the intersection of the vertical line and the extended graph line, draw a horizontal line to the distance axis. Then read the approximate distance of 240 kilometers.

▶ **Practice.**

1) Use the graph below. Find the approximate distance traveled for each of these times:

a) 2 hours   b) 3 hours
c) 4.5 hours   d) 2.5 hours
e) 8 hours   f) 5.5 hours
g) 7 hours   h) 10 hours

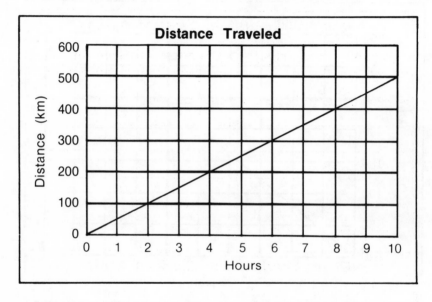

2) Make a line graph for the following table. On your graph use the same axis points that are on the graph above.

Elapsed Time (Hrs.)	Distance Traveled in km
1	100
2	200
3	300
4	400
5	500

When Speed Is Not Constant

Autos 1, 2, and 3 traveled at a constant speed. The lines for these autos on the graphs were straight.

Usually, however, an auto does not travel at a constant speed. Its speed changes from time to time. Look at the following table and graph. These show the motion of an auto whose speed has changed at various times.

Data Table for Non-Constant Speed	
Elapsed Time (Hrs.)	Distance Traveled in km
1	50
2	110
3	140
4	160
5	210
6	235

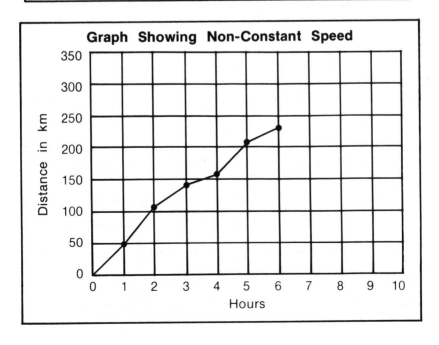

Graph Showing Non-Constant Speed

7.7 FORCE

Whenever any object changes its motion, a force causes the change. A *FORCE* is a push or a pull. For example, a car begins to move because of a force supplied by the engine. Look at the following examples of motion caused by a force:

7.8 ACCELERATION

When objects start from rest and begin to move, they are said to be accelerating. *ACCELERATION* means to change speed. The acceleration of an object can be calculated mathematically by using this formula:

$$\text{Acceleration} = \frac{\text{Change in speed}}{\text{Change in time}}$$

Look at the following example:

Example: Find the acceleration of an automobile that starts from rest at an intersection and has a speed of 40 km/hr after 5 seconds.

Step 1: Find the change in speed. Subtract the beginning speed from the final speed.

40	km/h	Final speed
− 0	km/h	Original speed
40	km/h	Change in speed

Step 2: Divide the change in speed by the time required to make the change.

The result is the acceleration.

$$\text{Acceleration} = \frac{\text{Change in speed}}{\text{Change in time}}$$

$$\text{Acceleration} = \frac{40 \text{ km}}{1 \text{ hr}} \div \frac{5 \text{ seconds}}{1}$$

$$\text{Acceleration} = \frac{40 \text{ km}}{1 \text{ hr}} \times \frac{1}{5 \text{ seconds}}$$

$$\text{Acceleration} = \frac{40 \text{ km}}{1 \text{ hr} \times 5 \text{ seconds}}$$

$$\text{Acceleration} = 8 \text{ km/hr per sec.}$$

The answer is read as 8 km per hour per second. This means that the auto's speed was increasing by 8 km per hour every second. Here is a table summarizing the auto's speed at each second.

Elapsed Time	Speed	Acceleration
0 sec.	0 km/hr	0 km/hr per sec.
1 sec.	8 km/hr	8 km/hr per sec.
2 sec.	16 km/hr	8 km/hr per sec.
3 sec.	24 km/hr	8 km/hr per sec.
4 sec.	32 km/hr	8 km/hr per sec.
5 sec.	40 km/hr	8 km/hr per sec.

Example: Find the acceleration of an automobile that begins with a speed of 40 km/hr and has a speed of 46 km/hr after 3 seconds.

Step 1: Find the change in speed by subtracting the beginning speed from the final speed.

$$
\begin{array}{ll}
46 \text{ km/hr} & \text{Final speed} \\
-40 \text{ km/hr} & \text{Original speed} \\
\hline
6 \text{ km/hr} & \text{Change in speed}
\end{array}
$$

Step 2: Divide the change in speed by the time required to make the change. The result is the acceleration.

$$\text{Acceleration} = \frac{\text{Change in speed}}{\text{Change in time}}$$

$$\text{Acceleration} = \frac{6 \text{ km}}{1 \text{ hr}} \div \frac{3 \text{ seconds}}{1}$$

$$\text{Acceleration} = \frac{6 \text{ km}}{1 \text{ hr}} \times \frac{1}{3 \text{ seconds}}$$

$$\text{Acceleration} = \frac{6 \text{ km}}{1 \times 3 \times \text{hr} \times \text{sec.}}$$

$$\text{Acceleration} = 2 \text{ km/hr per sec.}$$

The answer is read as 2 km per hour per second. This means that the auto's speed was increasing by 2 km per hour every second.

▶ **Practice.** Find the acceleration for each problem.

1) Beginning speed is 40 km/hr.
   Ending speed is 50 km/hr.
   Elapsed time is 5 sec.
   Acceleration = ?

2) Beginning speed is 20 km/hr.
   Ending speed is 55 km/hr.
   Elapsed time is 7 sec.
   Acceleration = ?

3) Beginning speed is 0 m/sec.
   Ending speed is 10 m/sec.
   Elapsed time is 10 sec.
   Acceleration = ?

4) Beginning speed is 20 mm/sec.
   Ending speed is 22 mm/sec.
   Elapsed time is .2 sec.
   Acceleration = ?

5) Beginning speed is 60 km/hr.
   Ending speed is 70 km/hr.
   Elapsed time is 2 sec.
   Acceleration = ?

6) Beginning speed is 30 m/min.
   Ending speed is 60 m/min.
   Elapsed time is 10 sec.
   Acceleration = ?

7) Beginning speed is 5 cm/sec.
   Ending speed is 10 cm/sec.
   Elapsed time is .5 sec.
   Acceleration = ?

8) Beginning speed is 37 km/sec.
   Ending speed is 77 km/sec.
   Elapsed time is 4 sec.
   Acceleration = ?

7.9 DECELERATION

You know that acceleration is a change in speed. The examples shown so far have been increases in speed. Objects can also slow down as a result of forces. When they slow down, they are said to *DECELERATE*. Deceleration is really negative acceleration.

In this next example, an auto is decelerating.

Example: An automobile is traveling at 20 km/hr. The driver suddenly puts on the brakes and comes to a complete stop 4 seconds later. Calculate the deceleration.

Step 1: The same formula is used as before.

$$\text{Deceleration} = \frac{\text{Change in speed}}{\text{Change in time}}$$

Step 2: The original speed was 20 km/hr. and the final speed is 0 km/hr. Therefore, the change in speed is −20 km/hr.

Step 3: Divide the change in speed by the change in time to obtain the deceleration.

$$\text{Deceleration} = \frac{-20 \text{ km}}{1 \text{ hr}} \div \frac{4 \text{ seconds}}{1}$$

$$\text{Deceleration} = \frac{-20 \text{ km}}{1 \text{ hr}} \times \frac{1}{4 \text{ seconds}}$$

$$\text{Deceleration} = \frac{-20 \text{ km}}{1 \text{ hr} \times 4 \text{ sec}}$$

$$\text{Deceleration} = -5 \text{ km/hr per sec.}$$

▶ *Practice.* Find the deceleration for each problem.

1) Beginning speed is 70 km/hr.
   Ending speed is 50 km/hr.
   Elapsed time is 5 sec.
   Deceleration = ?

2) Beginning speed is 80 km/hr.
   Ending speed is 55 km/hr.
   Elapsed time is 5 sec.
   Deceleration = ?

3) Beginning speed is 10 m/sec.
   Ending speed is 0 m/sec.
   Elapsed time is 10 sec.
   Deceleration = ?

4) Beginning speed is 30 mm/sec.
   Ending speed is 22 mm/sec.
   Elapsed time is .2 sec.
   Deceleration = ?

5) Beginning speed is 70 km/hr.
   Ending speed is 60 km/hr.
   Elapsed time is 2 sec.
   Deceleration = ?

6) Beginning speed is 25 m/min.
   Ending speed is 10 m/min.
   Elapsed time is 5 sec.
   Deceleration = ?

7) Beginning speed is 25 cm/sec.
   Ending speed is 10 cm/sec.
   Elapsed time is .5 sec.
   Deceleration = ?

8) Beginning speed is 20 km/sec.
   Ending speed is 10 km/sec.
   Elapsed time is 4 sec.
   Deceleration = ?

7.10 THE LAWS OF MOTION

Sir Isaac Newton was a brilliant scientist who studied the motion of bodies. He is responsible for formulating the *THREE LAWS OF MOTION*. Here are these three laws:

Law 1: Every object remains at rest or moves at constant speed in a straight line unless acted upon by an outside force.

Law 2: The acceleration of an object depends on the force acting on it and the mass of the object.

    a) The famous formula $F = ma$ is a mathematical statement of this law. It says that:

$$Force = mass \times acceleration$$

    b) The larger the force, the greater is the acceleration. The greater the mass, the lesser is the acceleration.

    c) In the metric system, force is measured in units called *NEWTONS*. A newton is the force needed to accelerate a mass of 1 kilogram by 1 meter/second per second.

    d) The unit of force in the English system is the *POUND*.

Law 3: For every force there is an equal and opposite force. This means that every action is accompanied by a reaction.

The First Law of Motion

The first law states that an object at rest (not moving) will remain at rest until a force is applied to it. In addition, the law says that an object in motion will remain in motion until a force acts on it. This seems to be against common sense.

If you throw a baseball, it certainly slows down, and, in fact, it comes to a stop. Likewise, an automobile will roll to a stop if your foot is kept off the gas pedal.

In both of these cases, invisible forces are at work. The baseball slows down because of the force of air resistance. The automobile slows down due to the force of *FRICTION.* Friction is a force that opposes motion.

If a baseball were thrown in outer space where there is no air, the ball would continue to move at constant speed until an outside force acted on it.

Because of friction, things that move tend to heat up. An auto engine heats up after it is started. To prevent it from overheating, the car has a radiator to get rid of the excess heat. In the case of automobile engines, friction is undesirable. Mechanics try to decrease friction by using oil and grease.

Some friction is good. When your hands are cold in the winter, you rub them together, and the heat makes them feel warmer.

Friction, then, can be considered either good or bad depending on the circumstances.

The Second Law of Motion

Which is easier to push — an empty wagon, or a wagon full of bricks? It takes less effort to push the empty wagon. The amount of force needed to move something depends on the mass of the object. The bigger the mass, the more force needed to move it.

A small force acting on a large mass will cause very little change in motion. A large force on a small mass will cause a much larger change in motion.

The Third Law of Motion

Imagine standing on a skateboard holding a large brick in your hand. If you throw the brick forward, you and the skateboard will move in the opposite direction from the brick. This is an example of action and reaction. Your *ACTION* of throwing the brick causes the *REACTION* of the skateboard moving backwards.

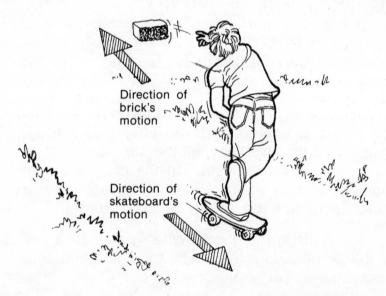

Direction of brick's motion

Direction of skateboard's motion

This picture illustrates Newton's third law of motion: For every force there is an equal and opposite force.

▶ **Practice.** Number your paper from 1 to 8. Write the answer to each question.

1) How many laws of motion are there?

2) Who formulated the laws of motion?

3) What is the unit of force in the metric system?

4) What is the unit of force in the English system?

5) When a marble is rolled along a floor, what force is responsible for causing it to slow down and stop?

6) State one example of when friction is useful.

7) State one example of when friction is not desirable.

8) If an object is at rest, what must happen for it to begin moving?

7.11 THE FORCE OF GRAVITY

One of the more important and familiar forces is the force of *GRAVITY*. Gravity is a force of attraction between any two objects that have mass. The gravitational force of a body depends on its mass. An object like the earth, which has a large mass, also has a large gravitational force. Smaller objects such as people, trees, and buildings have much smaller gravitational forces because they have less mass.

Universal Law of Gravitation

Sir Isaac Newton, who formulated the Three Laws of Motion, also formulated the Universal Law of Gravitation. That law says that any two objects attract each other with a gravitational force. The amount of that force depends on the mass of the two bodies and the distance between them. The larger the masses, the MORE is the gravitational force. The larger the distance between the objects, the LESS is the gravitational force.

Gravity and Weight

Because we live on the earth, the gravity of earth is important to us. It is because of the earth's gravity that we have weight. The earth attracts our bodies and our bodies attract the earth. It is really the force of gravity on our body which we measure when we weigh ourselves on a scale.

Acceleration Due to Gravity

Because gravity is a force, it can cause objects to change their motion. In other words, it causes acceleration. The letter "g" is used to stand for the acceleration due to gravity. The values for "g" in the metric and English systems are:

Metric System

g = 9.8 meters / second per second *or:*

g = 980 centimeters / second per second

English System

g = 32 feet / second per second

The speed of a falling object will increase because of the acceleration due to gravity. Each second its speed increases by 9.8 meters (or 32 feet) per second.

SUMMARY OF CHAPTER 7

1) Speed is the distance traveled per unit of time.

2) To calculate speed, divide the distance traveled by the time taken.

3) To find the distance an object travels, multiply the speed by the time.

4) To calculate the time an object takes to travel, divide the distance traveled by the speed.

5) A graph can be used to describe motion.

6) A constant rate of speed is shown on a graph by a straight line.

7) A distance-time graph can be used to predict distances.

8) A force is a push or a pull.

9) To accelerate means to change speed.

10) To decelerate means to slow down.

11) To calculate acceleration, divide the change in speed by the change in time.

12) Sir Isaac Newton formulated the Three Laws of Motion.

13) Friction is a force that opposes motion.

14) Gravity is the force of attraction between any two bodies which have mass.

15) The Universal Law of Gravitation states that all bodies attract each other.

16) The acceleration due to gravity is called "g."

• CHAPTER REVIEW EXERCISES •

Vocabulary Words

time	speed	distance
acceleration	average speed	constant speed
force	deceleration	Laws of Motion
friction	air resistance	action
reaction	Universal Law of	g
gravity	Gravitation	

▶ Copy the sentences. Fill in the missing words.

1) The distance divided by the ____ is the speed.
2) A push or a pull is called a ____.
3) For every action there is an equal and opposite ____.
4) Force is equal to mass times ____.
5) Sir Isaac Newton formulated three ____ of ____.

▶ Write the letter of the best ending for each sentence.

6) The formula for calculating distance is:
   a) $D = T \div S$ b) $D = S \times T$
   c) $D = S \div T$ d) $S = D \div T$

7) *Force = Mass times Acceleration* is part of the:
   a) First Law of Motion.
   b) Second Law of Motion.
   c) Third Law of Motion.

8) When an object slows down, it is said to be:
   a) accelerating.   b) reacting.
   c) decelerating.   d) acting.

9) Acceleration means a change in:
   a) distance.   b) time.   c) speed.   d) force.

10) An object travels 50 meters in 10 seconds. The speed of the object is:
   a) 60 m/sec.   b) 500 m/sec.
   c) 40 m/sec.   d) 5 m/sec.

Computer Programs

Type these programs into your computer to solve for the different components of motion — speed, distance, and time.

▶ *Solving for Speed:*

```
10  CLS
20  PRINT "ENTER THE DISTANCE"
30  INPUT D
40  PRINT "ENTER THE TIME"
50  INPUT T
60  PRINT "THE SPEED IS ";D/T
70  END
```

▶ *Solving for Distance:*

```
10  CLS
20  PRINT "ENTER THE SPEED"
30  INPUT S
40  PRINT "ENTER THE TIME"
50  INPUT T
60  PRINT "THE DISTANCE IS ";S * T
70  END
```

▶ *Solving for Time:*

```
10  CLS
20  PRINT "ENTER THE DISTANCE"
30  INPUT D
40  PRINT "ENTER THE SPEED"
50  INPUT S
60  PRINT "THE TIME IS ";D/S
70  END
```

Run each program. When you enter information, type the numerical portion only. Do not type units of measure, such as *miles* or *hours*. After you get an answer, attach the appropriate unit of measure to the answer.

Chapter **8**

Work and Machines

Chapter Goals

1. To calculate the amount of work done by a machine.

2. To describe the two basic kinds of energy.

3. To list six different types of energy.

4. To list and describe six different simple machines.

5. To calculate the work input and output of a lever.

6. To describe the difference between a fixed and movable pulley.

7. To calculate the mechanical advantage of a simple machine.

8.1 WORK

Have you ever been asked to do some "work" around the house? What things do you consider to be work? Ironing clothes, washing dishes, taking out the garbage, and sweeping the floors are common types of work around the house.

144

What do all of those household jobs have in common? In each case, something is moving, and a force is causing the motion. Scientists have a precise definition for work. *WORK* is a force moving something through a distance. Expressed in mathematical terms, *work* is defined as the force applied to an object multiplied by the distance the object moves.

WORK = FORCE APPLIED × DISTANCE MOVED

or

$$W = F \times d$$

where W means Work
       F means Force applied
       d means distance moved

Force is measured in *NEWTONS* in the metric system, and pounds in the English system. In either system a *SPRING SCALE* is used to measure force. A spring scale looks like the following diagram:

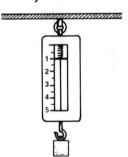

The usual units for work, force, and distance are shown in the following table.

Measurement	Metric System	English System
work	joules	foot-pounds
force	newtons	pounds
distance	meters	feet

Study the following examples of work problems.

Example 1: A girl pushes a bike using a force of 2 newtons. If she pushes the bike a distance of 10 meters, how much work will she do?

←——— 10 meters

Solution:

Work = Force × distance
W = F × d
W = 2 newtons × 10 meters
W = 20 newton-meters, or 20 joules

Example 2: A boy pulls a wagon 30 feet up a hill using a force of 10 pounds. How much work does he do?

10 lbs.
force →

30 feet →

Solution:

Work = Force × distance
W = F × d
W = 10 pounds × 30 feet
W = 300 foot-pounds

▶ **Practice.** Find the answer for each problem.

1) A man pushed a table using a force of 8 pounds. He moved the table 13 feet. How much work did he do?

2) A tractor pulls a trailer using a force of 500 newtons. How much work is done in pulling the trailer 10.5 meters?

3) A man lifts a box using a force of 10.5 pounds. If he lifts it 4 feet up, how much work is done?

4) The formula $W = F \times d$ can be solved for F:

$$F = W \div d$$

Use the formula to answer this question: How much force was needed to move an object 10 feet if the total work done was 100 foot-pounds?

5) A girl picked up her pet dog that weighs 15 pounds. She lifted the dog 4 feet up. How much work did she do?

6) A man slides a box up a ramp using a force of 32 pounds. If he slides the box a total of 120 feet, how much work does he do?

7) The formula $W = F \times d$ can be solved for d:

$$d = W \div F$$

Use the formula to answer this question: How high did a person lift a box if the force used was 50 pounds, and 100 foot-pounds of work was done?

8) What two quantities determine the amount of work done on an object?

9) What is the metric unit of work?

10) What is the English unit of work?

11) What is a force?

▶ Use this graph to answers questions 12 through 15.

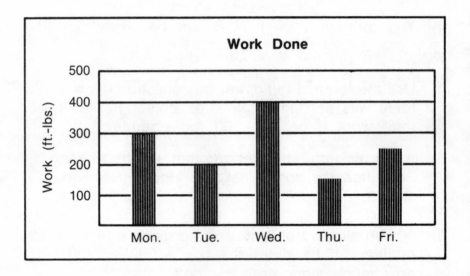

Work Done

12) On which day was the most work done?

13) How much work was done on Tuesday?

14) How much more work was done on Wednesday than on Tuesday?

15) How much work was done in all five days?

8.2 ENERGY AND WORK

Have you ever had one of those days when you said, "I just don't have any energy!"? Energy is

necessary to do work. In fact, *ENERGY* is defined as the ability to do work. Without energy, no work could be done at all.

Kinetic Energy

An object that is moving, such as a car, has energy which we call kinetic. *KINETIC ENERGY* is the energy of motion. The car can do work because it is moving. It has the ability to do work.

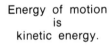

Energy of motion
is
kinetic energy.

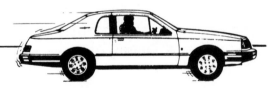

Potential Energy

There is another kind of energy that is not moving. Imagine a spring on a mousetrap. When the trap is set, it has energy stored in it. This energy can do work when the trap is released. The spring has *POTENTIAL ENERGY*. Potential energy is stored energy.

Stored energy
is
potential energy.

The Forms of Energy

There are six main forms of energy that we can use to do work. These are the six:

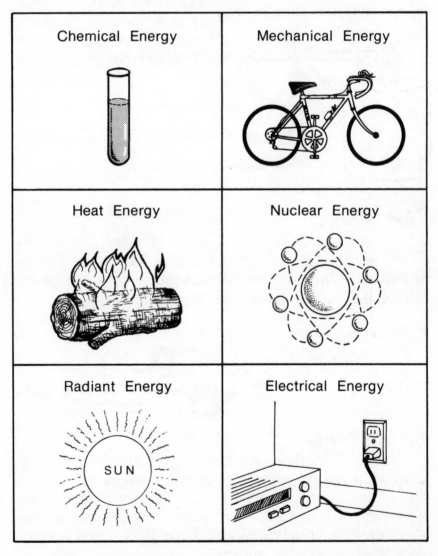

Chemical Energy	Mechanical Energy
Heat Energy	Nuclear Energy
Radiant Energy	Electrical Energy

Each of the forms of energy can do work. In our lives some forms of energy are more convenient to use than others. It is possible to change one form of energy into another.

Chemical energy, for example, in the form of coal, can be burned to make heat which changes water to steam. The steam can be made to turn a large generator. The generator makes electricity which we can use to run different appliances.

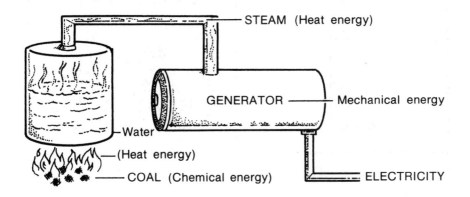

Because electricity can be used in such a variety of ways, most of the energy we use is in the form of electricity.

▶ **Practice.** Electrical appliances change electricity into another form of energy. For each appliance below, write the form of energy that the electricity is changed INTO. There may be more than one form.

1) toaster   2) light bulb
3) can opener   4) television
5) hair blower   6) clock

▶ Answer these questions on your paper.

7) What are six forms of energy?

8) What is the meaning of "energy"?

9) What is kinetic energy?

10) What device can change chemical energy into electrical energy?

11) Why is electricity a useful form of energy?

8.3 SIMPLE MACHINES

Have you ever tried to open a paint can with your fingers? It is difficult, if not impossible. A screwdriver, however, can help open the can quite easily with just a slight push. A screwdriver, when used in this way, is an example of a *SIMPLE MACHINE*.

Simple machines can do three things for us:

1) Increase our force.
2) Increase the speed of an object.
3) Change the direction of a force.

The Lever

One type of simple machine is the *LEVER*. A lever can have many shapes, but it commonly looks like this:

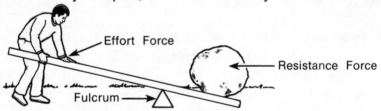

A simple machine such as a lever is used to increase your force, increase the speed of an object, or change the direction of a force. In the above case, the person pushes down on the lever while the object moves up. The force has changed its direction. In addition, it is much easier to lift the object using this lever. The force of the person has been increased.

The object that is to be moved is called the *RESIST-ANCE FORCE*. The force that the person applies to the lever is called the *EFFORT FORCE*. The object on which the lever rotates is called the *FULCRUM*.

The Three Classes of Levers

Levers can be classified into three groups depending on the position of the object, the fulcrum, and the person applying the force.

FIRST CLASS:

Here are some examples of first-class levers being used as tools.

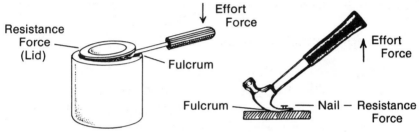

As you can see from the example of a hammer, a lever does not have to be a straight rod.

SECOND CLASS:

Here is a familiar example of a second-class lever.

THIRD CLASS: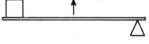

A broom is an example of a third-class lever.

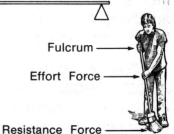

Work Done by a Lever

A simple machine such as a lever does not increase the amount of work done by a person. (Remember, in physics, *WORK* is the force multiplied by the distance.) Machines are usually used instead to increase the amount of effort; in other words, to make an object easier to lift.

Consider a person trying to use a lever to move a large boulder, as shown in this diagram:

The person can move the boulder with a small amount of force, but the boulder will only rise by a small distance compared to the distance you must push down on the lever.

The amount of work the person does in lifting the boulder is equal to the force times the distance moved. In this case the force is small, but the distance is large.

The other end of the lever has lifted a large force (the boulder), but it has only moved a small distance. This relationship is expressed mathematically as:

Work Input = Work Output

But the *WORK INPUT* is equal to the force multiplied by the distance, or:

Work Input = Effort Force × Effort Distance

The *WORK OUTPUT* is also equal to:

Work Output = Resistance Force × Resistance Distance

Here is a formula that shows how the elements are related:

$$F_E \times D_E = F_R \times D_R$$

Where: F_E is the EFFORT FORCE

D_E is the EFFORT DISTANCE

F_R is the RESISTANCE FORCE

D_R is the RESISTANCE DISTANCE

Study the following examples which use the work formula for levers.

Example 1:
A girl will use a lever to move a large box weighing 50 pounds.

The box is to be raised 1 foot up while the girl pushes down on the lever a distance of 2 feet. How much force must she use?

The problem asks us to find the *EFFORT FORCE*, or F_E.

FORMULA: $F_E \times D_E = F_R \times D_R$

$F_E \times 2$ ft. $= 50$ lbs. $\times 1$ ft.

$F_E \times 2$ ft. $= 50$ ft.-lbs.

$F_E = 50$ ft.-lbs. $\div 2$ ft.

$F_E = 25$ lbs. $\longleftarrow$ Answer

The girl need only push down with a force of 25 pounds to raise a box weighing 50 pounds! However, she must move the lever twice the distance that the box moves.

Example 2: A man will use a lever to move a boulder weighing 100 pounds. He needs to lift the boulder 1 foot high. If the man places the lever so that he needs to push down with only 50 pounds of force, how many feet will he move the lever? Find the *EFFORT DISTANCE*, or D_E.

FORMULA: $F_E \times D_E = F_R \times D_R$

$50 \text{ lbs.} \times D_E = 100 \text{ lbs.} \times 1 \text{ ft.}$

$50 \text{ lbs.} \times D_E = 100 \text{ ft.-lbs.}$

$D_E = 100 \text{ ft.-lbs.} \div 50 \text{ lbs.}$

$D_E = 2 \text{ feet}$ ← Answer

► ***Practice.***

1) Copy the chart below. Fill in the missing boxes by using the work formula for simple machines.

F_E	D_E	F_R	D_R
5 lbs.	6 feet	?	1 foot
20 lbs.	?	50 lbs.	2 feet
30 lbs.	?	45 lbs.	2 feet
10 lbs.	4 feet	20 lbs.	?
5 lbs.	?	100 lbs.	2 feet
16 lbs.	3 feet	?	1 foot
75 lbs.	2 feet	150 lbs.	?

2) A boy pushes down on a lever with a force of 15 pounds. He pushes the lever down a distance of 2 feet. A box weighing 60 pounds is lifted up by the other side of the lever. How high does it move?

Pulleys

Another type of simple machine is a *PULLEY*. A pulley is a type of wheel with a rope or string around it.

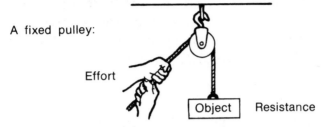

A fixed pulley:

Effort

Object   Resistance

A single pulley like the one above can be used to change the direction of your force. It does not increase your force as a lever does. This type of pulley can be used to lift a heavy object by pulling down instead of picking up.

The above pulley is called a *FIXED* pulley because it is fixed or attached at the top. It is not free to move up and down. It can only turn around.

Another type of pulley is a *MOVABLE* pulley.

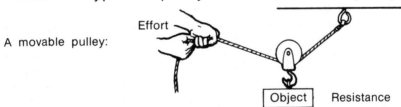

Effort

A movable pulley:

Object   Resistance

As effort is applied to this pulley, the entire pulley and the object attached will begin to rise. The rope supports the pulley from two directions. So the force needed to lift the object will be divided into two. This type of pulley, then, can be used to make a lifting job easier. Again, there is a price to pay for making it easier to lift — you must pull twice the distance to make the object move. For example, to lift the object 1 foot, you must pull up a distance of 2 feet on the rope.

Pulleys can be combined together to create different setups. Look at the following examples:

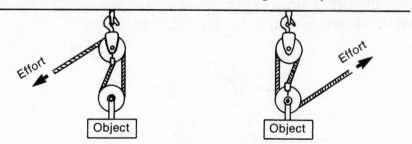

By experimenting with different pulley systems, you can design your own more complicated machines.

▶ *Practice.* Answer these questions.

1) What is a pulley?

2) What is a fixed pulley?

3) List two examples of where a single fixed pulley could be used.

4) What is a movable pulley?

5) In the following pulleys, how much effort force is necessary to lift the object?

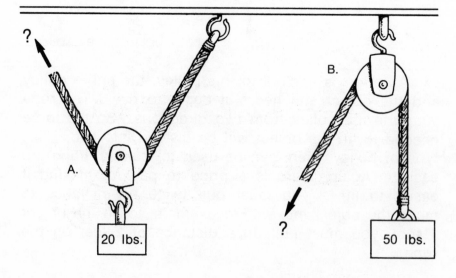

Inclined Planes

An unusual type of simple machine is the *INCLINED PLANE*. It is unusual because it has no moving parts. Here is an example of an inclined plane:

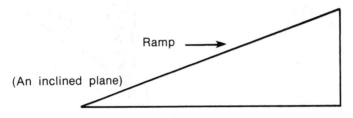

Ramp ⟶

(An inclined plane)

Another name for an inclined plane is the word *RAMP*.

Inclined planes *decrease the force* necessary to move an object but *increase the distance* it has to be moved. For example, if a delivery man needs to put boxes on a truck that is four feet from the ground, he might use an inclined plane or ramp to make his job easier. Rather than lifting each box up four feet, he can push each box up the ramp. It takes less force to push an object than to pick it up.

Other Simple Machines

In addition to the simple machines discussed so far, there are others. Another form of an inclined plane is a *SCREW*. A screw can be thought of as a straight piece of metal with an inclined plane wrapped around it.

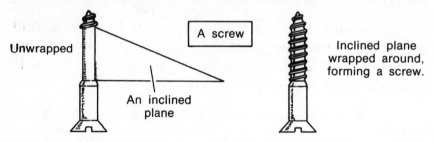

Screws make it much easier to fasten two objects together.

If two inclined planes are joined together, they can form a *WEDGE*. A wedge can be used for jobs like splitting logs apart.

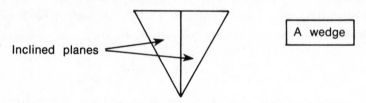

Another simple machine is a *WHEEL* and *AXLE*. In this machine, a *wheel* is attached to a shaft called an *axle*. A diagram of the wheel and axle is shown below.

A wheel and axle will increase the force from the wheel to turn something else attached to the axle. Examples of a wheel and axle are doorknobs and automobile steering wheels.

8.4 MECHANICAL ADVANTAGE

Nearly all of the simple machines are used to multiply the force which you apply. The number of times a machine multiplies your force is called the *MECHANICAL ADVANTAGE* of the machine.

For example, if you apply a force of 10 pounds, and the object lifted weighs 30 pounds, the mechanical advantage (abbreviated as MA) equals 3. This is because 30 ÷ 10 = 3.

Effort force is 10 pounds. Resistance force is 30 pounds.

Machine multiplies force 3 times.

To find mechanical advantage (MA), use the formula:

MA = RESISTANCE FORCE ÷ EFFORT FORCE

$$MA = F_R \div F_E$$

Example 1: Find the MA of a machine that lifts an object weighing 75 pounds, using an effort force of 15 pounds.

MA = RESISTANCE FORCE ÷ EFFORT FORCE

$MA = F_R \div F_E$

MA = 75 pounds ÷ 15 pounds

MA = 5 ◄——— Answer

The MA has no units. It is a pure number that tells how many times a force is multiplied.

Example 2: Find the MA of a machine that lifts an object weighing 100 pounds, using an effort force of 9 pounds.

$$MA = F_R \div F_E$$

$$MA = 100 \text{ pounds} \div 9 \text{ pounds}$$

$$MA = 11.1 \longleftarrow \text{ Answer}$$

SUMMARY OF CHAPTER 8

1) Work is defined as a force which moves an object through a distance.

2) The amount of work done is equal to the force multiplied by the distance.

$$\text{Work} = \text{Force} \times \text{Distance}$$

3) The metric unit of force is the *newton*.

4) The English unit of force is the *pound*.

5) The metric unit of work is the *joule*.

6) The English unit of work is the *foot-pound*.

7) Forces are measured using a spring scale.

8) Energy is the ability to do work.

9) The energy of motion is called *kinetic energy*.

10) Stored energy is called *potential energy*.

11) The six forms of energy are: chemical, mechanical, heat, nuclear, radiant, and electrical.

12) Electrical energy is the most convenient form of energy.

13) Simple machines can increase force, change the direction of a force, or increase the speed of an object.

14) A *lever* is a simple machine that has a fulcrum.

15) A *fulcrum* is the point on which a lever rests.

16) There are three classes of levers.

17) For a simple machine, the work input is equal to the work output.

18) The two types of pulleys are *fixed* and *movable*.

19) Inclined planes, wedges, screws, and a wheel and axle are other types of simple machines.

20) Mechanical advantage is a number that tells how many times a machine multiplies your force.

21) The MA of a machine is equal to the *resistance force* divided by the *effort force*.

Vocabulary Words

work
force
distance
spring scale
joule
foot-pound
newton
pound
energy
potential energy
kinetic energy
chemical energy
mechanical energy
heat energy
nuclear energy
radiant energy

electrical energy
simple machine
lever
fulcrum
effort force
resistance force
pulley
fixed pulley
movable pulley
inclined plane
ramp
screw
wedge
wheel and axle
mechanical advantage

• CHAPTER REVIEW EXERCISES •

▶ Copy the sentences. Fill in the missing words.

1) The number of times a machine multiplies your force is called the ___.

2) ___ is the ability to do work.

3) Work is measured in ___ in the metric system.

4) A type of pulley which is attached to something and cannot move is called a ___ ___.

5) Another name for an inclined plane is a ___.

▶ Write the letter of the best ending for each sentence.

6) Work can be measured in:
   a) pounds.   b) joules.   c) newtons.

7) A resting point for a lever is called a
   a) fulcrum.   b) ramp.   c) force.

8) The energy of motion is called
   a) potential.   b) nuclear.   c) kinetic.

9) A force of 10 pounds moves a distance of 10 feet. The work done is:
   a) 100 ft.-lbs.   b) 1 ft.-lb.   c) 20 ft.-lbs.

10) A machine formed by two inclined planes joined together is a
    a) screw.   b) wedge.   c) wheel and axle.

Calculator Practice

▶ Use your calculator to find the work done.

1) force = 9.86 newtons
   distance = 6.09 m

2) force = 18.392 pounds
   distance = 4.032 ft.

▶ Use your calculator to find the missing quantity. Use the formula $F_E \times D_E = F_R \times D_R$.

3) F_E = 36.02 pounds
   D_E = 18.93 feet
   F_R = 76.93 pounds
   D_R = ?

4) F_E = ?
   D_E = 18.69 feet
   F_R = 66.23 pounds
   D_R = 3.02 feet

5) F_E = 123.06 pounds
   D_E = 33.22 feet
   F_R = 486.39 pounds
   D_R = ?

6) F_E = 82.66 pounds
   D_E = 3.98 feet
   F_R = ?
   D_R = 26.2 feet

Computer Program

▶ Use this computer program to calculate the mechanical advantage of a machine.

```
10 CLS
20 PRINT "ENTER THE EFFORT FORCE ";
30 INPUT EF
40 PRINT:PRINT
50 PRINT "ENTER THE RESISTANCE FORCE ";
60 INPUT RF
70 PRINT:PRINT
80 LET MA=RF/EF
90 PRINT "THE MA= ";MA
100 PRINT:PRINT
110 PRINT "WANT TO DO ANOTHER? (Y/N) ";
120 INPUT R$
130 IF R$="Y" THEN 10
140 CLS
150 END
```

Chapter 9

Heat

Chapter Goals:

1. To describe four ways heat can be produced.

2. To list the three states of matter.

3. To define the word *temperature*.

4. To explain how a thermometer measures temperature.

5. To define the properties of boiling point, melting point, and freezing point.

6. To calculate heat gained or lost.

7. To state the three methods by which heat travels.

Is there anything more inviting than a roaring fire in a fireplace on a cold winter evening? The fire gives us light and warms our bodies when we're chilled.

9.1 WHAT IS HEAT?

As you watch a fire and feel its heat, a natural question to ask is, "What is heat?" That question can be answered by saying that heat is a form of energy made by the motion of molecules. It is energy because it has the ability to do work. Heat energy is used to make steam engines. Heat energy also turns generators that make electricity.

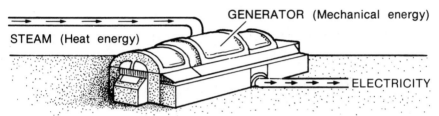

GENERATOR (Mechanical energy)

STEAM (Heat energy)

ELECTRICITY

9.2 HOW IS HEAT ENERGY PRODUCED?

Heat is a form of energy. Other forms of energy can be changed into heat energy. Rub your hands together rapidly. Do they feel warm? The heat produced by motion (friction) can be felt.

Heat can also be produced by electricity. A toaster is an example of an appliance that changes electricity into heat energy. Can you think of others?

Another source of heat is chemical energy. Some chemicals can be made to give off heat. Natural gas, for example, will burn and give off heat.

Nuclear energy can also be changed into heat energy. Our most important source of heat energy, the sun, makes its heat by nuclear reactions.

SUN

The earth's most important source of heat energy

Without the sun's source of heat energy, all life on earth would quickly die out.

9.3 WHAT ARE THE EFFECTS OF HEAT?

The effects of heat can be seen around us all the time. When you boil water on the stove, the water is being heated. Eventually, the water will be gone or *EVAPORATED*. A liquid evaporates when it changes into a gas because of heat.

In the summertime, if you leave an ice cream cone in the heat of the sun, you know that it will begin to melt or change into a liquid. Again, the change from a solid to a liquid is due to heat.

9.4 THE THREE STATES OF MATTER

Matter can be found in any one of three different *STATES*. The three states of matter are *SOLIDS*, *LIQUIDS*, and *GASES*.

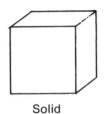
Solid

Liquid

Gas

The states of matter are related to heat. If sufficient heat is added to a solid, the solid will eventually begin to melt and change into a liquid. If the liquid is heated further, it will change into the gas state.

The most familiar substance which shows all three states of matter is water. Water in the solid state is called *ICE*. In the liquid state it is called *WATER*, and in the gas state it is called *STEAM*. The state of matter depends on the temperature of the substance.

Solid   Liquid   Gas

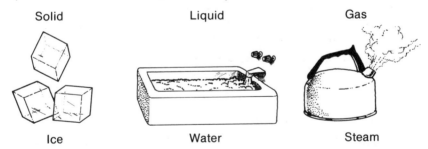

Ice   Water   Steam

9.5 WHAT IS TEMPERATURE?

All substances are made of molecules or atoms. These molecules and atoms are constantly in motion. Scientists measure how fast the molecules are moving by using a *THERMOMETER*. A thermometer is a measuring device used to measure temperature. The faster the molecules move, the higher is the temperature.

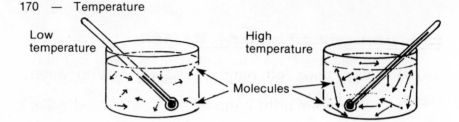

Low temperature

High temperature

Molecules

The scientist's definition of temperature seems to be quite different from what the normal experience of temperature is. People certainly do not *feel* molecules moving faster when an object is hot. We describe temperatures as either hot, warm, cool, or cold.

Scientists cannot depend on humans' descriptions of temperature. As a measuring device, people are not accurate. Try the following experiment.

INVESTIGATION 6

Purpose: To show that human descriptions of temperature are not accurate.

A. Materials

Quantity:	*Item Description:*
3	Pans or large jars
5	Ice cubes

B. Procedure

1) Fill one of the three pans with very warm tap water.

2) Fill one of the three pans with lukewarm tap water.

3) Fill the remaining pan with cold tap water. If possible, add some ice cubes to the water.

4) Place your left hand in the very warm water.

5) Place your right hand in the pan of cold water.

6) Leave both hands in the pans for at least 15 seconds.

7) Take both hands out of the pans at the same time. Then place both your hands in the lukewarm water.

Cold   Lukewarm   Hot

C. Results

What temperatures did you feel with your left hand? What temperatures did you feel with your right hand? On a separate sheet of paper, describe the temperatures you felt with each hand.

The results of the previous investigation show that the hand that was in the warm water will feel cold when placed in the lukewarm water. The hand that was in the cold water will feel warm when placed in the lukewarm water. The lukewarm water is at one temperature, but our descriptions of that temperature can be quite different. Because of these inaccurate descriptions, scientists use special instruments rather than rely on their own descriptions.

▶ *Practice*. Write the answers.

1) Name three sources of heat energy.

2) What is evaporation?

3) What is the earth's most important source of heat?

4) Name the three states of matter.

5) What instrument is used to measure temperature?

6) What is temperature?

7) Why do scientists use thermometers instead of human descriptions of temperature?

8) Classify each of the following as a solid, a liquid, or a gas.

   a) milk   b) sand

   c) steam   d) orange juice

   e) air   f) wood

   g) saliva   h) ice

   i) rock   j) soup

   k) oil   l) helium

   m) paper   n) root beer

9) What happens to the temperature of an object as the molecules move faster?

10) What is the sun's source of heat energy?

11) What are the three forms of water called?

9.6 MEASURING TEMPERATURE

Scientists use thermometers to measure temperature. When a substance becomes hotter, the molecules move faster and faster. This increased movement causes the molecules to move further apart. When the molecules of a substance move further apart, we say that the substance is *EXPANDING*, or becoming bigger.

In a mercury thermometer, a glass tube is filled with liquid mercury. The ends of the tube are closed off. As the thermometer measures higher temperatures, the mercury molecules begin to move further apart. The liquid expands. The only direction in which the mercury can move is *up* the tube. By placing markings on the tube, we can assign numbers to the height of the mercury.

The mercury expands and moves up the tube

←—Heat

Not all thermometers are made from mercury. Some inexpensive thermometers are filled with alcohol which has red dye added to it so it is easier to read.

9.7 FAHRENHEIT AND CELSIUS TEMPERATURE SCALES

There are two common scales used to measure temperatures. People in the United States use the *FAHRENHEIT* scale. People in most of the other countries of the world use the *CELSIUS* scale. The *CELSIUS* scale is also used by scientists.

On the next page you can see a comparison of the two scales.

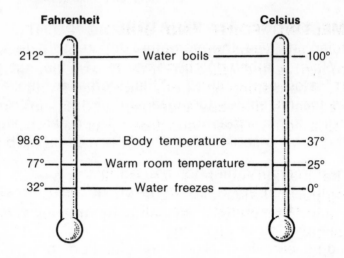

Most people in the United States are familiar with the Fahrenheit scale. The Celsius scale, however, is easier to use and remember because of the temperatures of freezing water (0°), and boiling water (100°). Regardless of the scale being used, the numbers are always read as "degrees." Then the name of the scale is given. The symbol for degrees is °. Look at the following examples.

Temperature:	Read as:
a) 87°F	87 degrees Fahrenheit
b) 32°C	32 degrees Celsius
c) 18°C	18 degrees Celsius
d) 36°F	36 degrees Fahrenheit
e) −10°F	Minus 10 degrees Fahrenheit, or 10 degrees below zero Fahrenheit
f) − 5°C	Minus 5 degrees Celsius or 5 degrees below zero Celsius

9.8 MELTING POINT AND BOILING POINT

Two important temperatures for all substances are known as the *MELTING POINT* and the *BOILING POINT*. The melting point of a substance is the temperature at which the substance begins to change from a solid to a liquid. Every substance has its own melting point.

The melting point of ice is 32°F or 0°C. The melting point of alcohol is −202°F or −130°C. There is quite a difference in the melting points of different substances!

The term *melting point* is usually used when a substance is being heated. If the substance is being cooled from a liquid to a solid, the term *FREEZING POINT* would be used. The freezing point and melting point are the same temperature for any one substance.

The Boiling Point

The *BOILING POINT* of a substance is the temperature at which the substance changes from a liquid to a gas. For example, water boils at 212°F or 100°C. Therefore, the boiling point of water is 212°F or 100°C. The boiling point of alcohol is 173°F or 78°C. Liquid alcohol would change to a gas at that temperature.

Melting and Boiling Points of Various Substances					
Melting Point				Boiling Point	
Fahrenheit	Celsius	Substance	Fahrenheit	Celsius
32°	0°	Water	212°	100°
1220°	660°	Aluminum	4473°	2467°
1764°	962°	Silver	4014°	2212°
−38°	−39°	Mercury	675°	357°
−202°	−130°	Alcohol	173°	78°

9.9 BOILING POINT AND FREEZING POINT CAN BE CHANGED

Although all substances have a certain boiling point and freezing point, those temperatures can be changed by mixing substances together. For example, by adding alcohol to water, the freezing point will be lower than 32°F. This is what is done when antifreeze is added to the radiator in an automobile. On a cold winter

evening, if the water in your car's engine had no antifreeze, the water would freeze, damaging the engine.

Also in the winter, rock salt is often thrown on ice to remove it from sidewalks. Rock salt lowers the freezing point of water. It causes the ice to turn back into a liquid. If the temperature is too cold, however, the salt will not work very well, since it only lowers the freezing point by a few degrees.

9.10 HEAT AND TEMPERATURE

Although we sometimes use the words *temperature* and *heat* interchangeably, the two words do not mean the same thing to the scientist. Temperature is a measure of how fast the molecules are moving in a substance. It is measured in units called degrees.

Heat, on the other hand, depends on the temperature AND the amount of matter (*MASS*) which is present. As the temperature of an object increases, the amount of heat in the object is increased also. Another way of looking at this is that if two objects of different mass are at the same temperature, the larger object will contain more heat.

An example can help to make this clearer. The temperature of a match and a bonfire can be measured using a thermometer and found to be at the same temperature (assuming that they are both made of the same material). However, the bonfire contains more heat than the match. The bonfire will keep you warm much longer than a match because the bonfire has more mass.

Match

Bonfire

Scientists measure heat in units called *CALORIES*. A calorie is the amount of heat needed to raise the temperature of 1 gram of water, 1 degree Celsius.

> 1 Calorie = The amount of heat needed to raise the temperature of 1 gram of water by 1 degree Celsius

Two calories would raise the temperature of 1 gram of water by 2 degrees Celsius. This relationship is expressed mathematically as:

Calories = Change in temperature × number of grams of water

Study the following example problems:

1) How many calories of heat are needed to raise the temperature of 1 gram of water by 3 degrees C?

   Method:

   Calories = change in temperature × grams of water

   Calories = 3 degrees × 1 gram

   Calories = 3

2) How many calories of heat are needed to raise the temperature of 5 grams of water by 10 degrees C?

Method:

Calories = change in temperature × grams of water

Calories = 10 degrees × 5 grams

Calories = 50

3) How many calories of heat are needed to raise the temperature of 6 grams of water from 5° Celsius to 15° Celsius?

Method: First calculate the temperature change.
Change in temperature = 15°C − 5°C
= 10°C
Now calculate the heat.

Calories = change in temperature × grams of water

Calories = 10 degrees × 6 grams

Calories = 60

▶ *Practice.* Write the answers to these questions.

1) 25 grams of water are heated 5°C. How many calories are needed?

2) The temperature of 8 grams of water is raised 8 degrees Celsius. How much heat is added?

3) 15 grams of water are heated from 12°C to 22°C. How many calories are needed?

4) If 20 grams of water are heated 1°C, how much heat is added?

5) If 35 grams of water are heated from 10°C to 11°C, how much heat is added?

Cooling

Heat can also be lost or given off by water. This happens when the water is cooling. To indicate that the water is being cooled, a minus sign (−) is placed in front of the answer. Study the following problem.

Example: 20 grams of water are cooled from 20°C to 10°C. How much heat is given off?

First calculate the temperature change:

Temperature change = 20°C − 10°C

Temperature change = 10°C

Now calculate the calories:

Calories = change in temperature × grams of water

Calories = 10 degrees × 20 grams

Calories = −200

▶ *Practice.*

1) 35 grams of water are cooled from 15°C to 10°C. How much heat is given off?

2) 15 grams of water are cooled from 100°C to 25°C. How much heat is given off?

3) 50 grams of water are cooled from 25°C to 24°C. How much heat is given off?

4) 25 grams of water are cooled from 18°C to 16°C. How much heat is given off?

5) 88 grams of water are cooled from 22°C to 16°C. How much heat is given off?

6) 16 grams of water are cooled from 13°C to 1°C. How much heat is given off?

9.11 HOW DOES HEAT TRAVEL?

We know that the sun heats the earth. But, how does the sun's heat get from its surface to the earth?

Somehow the heat of the sun must travel a long distance to get to the earth. In addition, the heat must travel through a *VACUUM*. A vacuum means that there are no particles of matter, either molecules or atoms.

When heat travels through a vacuum, it moves as electromagnetic waves. These waves are similar to radio waves which also can travel through a vacuum. This type of heat movement is known as *RADIATION*.

Heat waves traveling by radiation are known as *INFRARED* waves. You can feel infrared heat waves when you sit in the sunshine.

Heat can also travel from one place to another by *CONDUCTION*. In the following diagram a piece of copper is being heated by a flame.

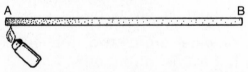

As heat is added to the end labeled A, the molecules near point A begin to move more rapidly. As they move more rapidly, they hit other molecules further away from A. When the molecules bump into each other, they cause other molecules to move faster. Eventually, after more heat is added, all the molecules in the copper are in rapid motion. In other words, the entire piece of copper becomes hot. Heat moves by conduction when the molecules strike other molecules and cause them to move more rapidly.

Some materials allow molecules to move more readily than others. Most metals, for example, will do this. Any substance that will allow heat to flow through it easily is called a *CONDUCTOR*.

Other materials will not allow heat to move easily. You can hold a wooden stick in the flames of a fire and not feel the heat in your hands at all. Wood is a poor conductor of heat. Just because wood will burn does not mean it is a good conductor of heat! Any substance that does not conduct heat well is called an *INSULATOR*.

Firefighters use special suits made from materials which are good insulators. In other words, the materials are poor conductors of heat. In a fire, the material of the suit does not allow the heat to travel inside the suit, thereby keeping the person safe.

Here is a list of materials that are conductors and insulators.

Insulators	Conductors
glass	steel
wood	copper
dirt	silver
sand	gold
asbestos	aluminum
styrofoam	tin

A third method of heat movement is called *CON-VECTION*. Convection occurs when part of a gas or liquid becomes heated. As the liquid or gas becomes hotter, it becomes less dense and begins to rise. Because of this rising, the gas or liquid begins to circulate. The air in a house begins to circulate because of convection.

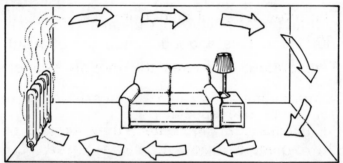

This transfer of heat by convection helps to keep the rooms in your house warm by causing all the air to move. In that way, the whole room becomes heated and not just a small part of the room.

▶ *Practice.* Write the answers.

1) Name the three methods by which heat travels.

2) How does the sun's heat travel to earth?

3) What is a vacuum?

4) When a pan is heated on a stove, by what method is the heat traveling?

5) Many good jackets are filled with goose down. Is goose down an insulator or a conductor of heat?

6) Name two substances that are insulators.

7) Name two substances that are conductors.

8) Define the word *conductor*.

SUMMARY OF CHAPTER 9

1) Heat is a form of energy.

2) Heat can be produced by electrical, nuclear, chemical, and mechanical energy.

3) Evaporate means to change from a liquid to a gas.

4) The three states of matter are *SOLID*, *LIQUID*, and *GAS*.

5) Temperature measures the motion of molecules.

6) A thermometer is used to measure temperature.

7) As substances are heated, they expand because the molecules begin to move faster.

8) The two commonly used temperature scales are the Fahrenheit and Celsius scales.

9) The temperature at which a solid changes into a liquid is the melting point.

10) The temperature at which a liquid changes into a gas is called the boiling point.

11) The freezing point of water is 32°F or 0°C. The boiling point of water is 212°F or 100°C.

12) Boiling points and freezing points can be changed by mixing substances together.

13) Heat depends on the temperature and the mass of an object.

14) Heat is measured in calories.

15) The number of calories gained or lost by water is equal to the change in Celsius temperature multiplied by the mass.

16) One calorie is the amount of heat needed to raise the temperature of 1 gram of water by 1 degree C.

17) Heat can travel by radiation, conduction, and convection.

18) Objects which allow heat to travel easily are called *conductors*, while those which do not are called *insulators*.

Temperature Conversion Table					
°C		°F	°C		°F
100	—	212	45	—	113
95	—	203	40	—	104
90	—	194	35	—	95
85	—	185	30	—	86
80	—	176	25	—	77
75	—	167	20	—	68
70	—	158	15	—	59
65	—	149	10	—	50
60	—	140	5	—	41
55	—	131	0	—	32
50	—	122			

Vocabulary Words

heat	thermometer	calorie
solid	expand	vacuum
liquid	Fahrenheit	radiation
evaporate	Celsius	conduction
gas	boiling point	convection
states of matter	melting point	insulator
temperature	freezing point	conductor

• CHAPTER REVIEW EXERCISES •

▶ Copy the sentences. Fill in the missing words.

1) As substances are heated, they get bigger, or ____.

2) A ____ is used to measure temperature.

3) An object which allows heat to travel through it easily is called a ____.

4) Heat is measured in units called ____.

5) Water boils at ____ degrees Fahrenheit.

▶ Write the letter of the best answer for each item.

6) Which of the following is NOT a method by which heat travels?
   a) radiation   b) expand   c) convection

7) The temperature at which a liquid changes into a solid is called the:
   a) boiling point.
   b) freezing point.
   c) melting point.

8) When heat warms a room by circulating air, heat is moving by:
   a) convection.   b) conduction.   c) radiation.

9) Water freezes at 0°:
   a) heat.   b) Fahrenheit.   c) Celsius.

10) Which of the following is not a state of matter?
    a) gas   b) solid   c) calorie

Calculator Practice

► Calculate the number of calories gained. Use your calculator.

	Temperature Change	Mass	Calories
1)	18.5°C	65.36 g	?
2)	13.2°C	81.009 g	?
3)	22.9°C	12.29 g	?
4)	13°C	6.22 g	?
5)	82.1°C	1.09 g	?
6)	19.2°C	16.03 g	?

Computer Program

► The following computer program will calculate the number of calories of heat gained or lost by a substance. You must first calculate the temperature change. Subtract the beginning temperature and the final temperature. Remember, if the temperature goes down, type in a minus sign before the number for the temperature change.

```
10 CLS
20 PRINT "THIS PROGRAM CALCULATES HEAT"
30 PRINT "GAINED OR LOST"
40 PRINT:PRINT
50 INPUT "WHAT IS THE MASS ";M
60 PRINT:PRINT
70 INPUT "WHAT IS THE TEMPERATURE CHANGE ";T
80 CLS
90 LET H=M * T
100 PRINT "THE ANSWER IS ";H;" CALORIES"
```

Chapter **10**

Sound and Light

Chapter Goals:

1. To state the cause of sound waves.

2. To explain how sound travels.

3. To define *decibel*, *pitch*, *frequency*, *vibration*, and *cycle*.

4. To describe a method of determining underwater distances.

5. To construct a tin-can telephone.

6. To name the colors of the spectrum.

7. To define *reflection* and *refraction*.

8. To list the two types of images.

9. To name three kinds of mirrors and two kinds of lenses.

10.1 INTRODUCTION

Rock musicians use a lot of energy creating their music on stage. What you may not realize is that their music itself is energy. This sound energy is the subject of the branch of science known as *ACOUSTICS*.

10.2 SOUND IS CAUSED BY VIBRATION

Whenever an object *VIBRATES*, it creates a movement in the air molecules surrounding it. To vibrate means to move back and forth. The movement of the air molecules caused by vibration is a *SOUND WAVE*.

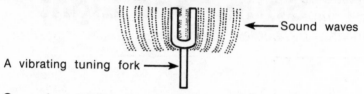

Sound waves

A vibrating tuning fork ——→

Sound waves travel outward in all directions from a vibrating object. As the sound waves travel further and further from the object, they become weaker.

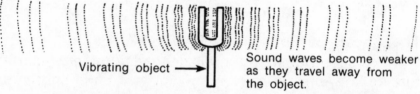

Vibrating object ——→

Sound waves become weaker as they travel away from the object.

10.3 SOUND TRAVELS THROUGH DIFFERENT SUBSTANCES

For sound to travel, there must be molecules which can be made to vibrate. Without any molecules, sound cannot travel at all. In outer space, for example, there are no molecules, and therefore no sounds. What a strange place it would be with no sound at all!

Sound travels through various substances by causing the molecules to vibrate and bump into other molecules. This process is similar to the action of a group of dominoes set up so that each domino hits the next one, causing it to fall over.

Each domino strikes the next, knocking ——→ all of them down.

Each of the dominoes travels only a short distance, but the effect of each domino's motion can travel large distances.

If you put the dominoes closer together, it will take less time to knock them all down. In the same way, sound will travel through substances more quickly when the molecules are packed closer together.

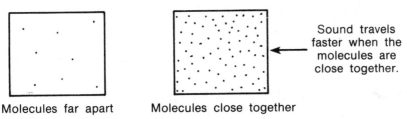

Sound travels faster when the molecules are close together.

Molecules far apart   Molecules close together

The Speed of Sound

The speed of sound in air depends on the temperature of the air. A convenient figure for the speed of sound in air is 346 meters per second at 25 degrees Celsius. This speed is about equal to 700 miles per hour. The speed of sound in some other materials is shown in the following table.

Speed of Sound in Various Materials at 25°C	
Material	Speed (Meters per second)
Air	346
Water	1,497
Wood	1,850
Steel	5,200

From the above table, you can see that sound travels much faster through steel than through air. Sound also travels faster through water than through air. The next time you are swimming, strike two rocks together underwater and see how well the sound travels through water.

▶ *Practice.* Answer these questions.

1) What causes sound waves?
2) What is the meaning of the word *vibrate*?
3) What happens to the speed of sound as the molecules in a substance are further apart?
4) What is the speed of sound in air in miles per hour?
5) What is the science of sound?

10.4 THE LOUDNESS OF SOUND

You know that some sounds are louder than others. You also know the further you are from the source of a sound, the softer the sound becomes. Scientists use the *DECIBEL* to describe the intensity of sounds. For example, a faint whisper would be measured as 20 decibels. The sound due to heavy traffic is about 80 decibels. The sound of a jet aircraft at takeoff is approximately 135 decibels.

A whisper is
about 20 decibels.

A jet engine is
about 135 decibels.

Each increase of 10 decibels means that the intensity of a sound you hear is twice as loud. For example, a sound with a decibel level of 20 decibels is twice as intense as a sound of 10 decibels. Our ears interpret the intensity as loudness. The more intense a sound, the louder we hear it.

10.5 FREQUENCY AND PITCH

The *FREQUENCY* of a sound wave describes the number of times the sound wave vibrates in one second. For example, if an object vibrates 5 times in each second, the resulting sound wave would have a frequency of 5 cycles per second. A *CYCLE* means one complete back-and-forth motion.

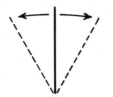

One cycle means one complete back-and-forth motion.

Frequency is the word which describes the number of vibrations per second of a sound wave. Although objects can vibrate at many different rates, the human ear can only hear a certain range of frequencies. The range depends on the conditions of the person's ears. Generally, the human ear can detect sounds with frequencies ranging from 20 cycles per second up to 20,000 cycles per second.

The sounds we hear are not usually described in terms of their frequency. Rather, we use the word *PITCH* to describe how sounds of different frequencies are heard by the ear. A high frequency sound wave has a high pitch, like that of a violin.

High frequency = High pitch

A low frequency sound wave has a low pitch, perhaps like that of a tuba. Your voice changes from a low pitch to higher pitches when you sing the musical scale:

do   re   mi   fa   sol   la   ti   do

▶ **Practice.** Answer these questions.

1) What is frequency?

2) What is the difference between frequency and pitch?

3) What unit is used to measure the loudness of sounds?

4) If a sound of 30 decibels gets twice as loud, how many decibels is the new sound?

5) Sound wave A has a frequency of 10,000 cycles per second. Sound wave B has a frequency of 15,000 cycles per second. Which wave has the higher pitch?

10.6 SOUND WAVES ARE REFLECTED

Look at yourself in a mirror. What you see is a *REFLECTION* of yourself. The image of your face is reflected back to your eyes. Sound waves can also be reflected, but they do not require a mirror. Sounds can be reflected by almost all objects.

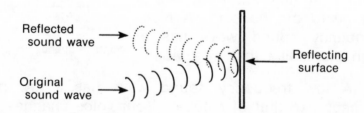

When a sound wave is reflected back to the original source of the sound, it is called an ECHO. You have probably heard an echo of your own voice many times.

10.7 USING ECHOES

SONAR is a device which measures distances underwater by sound waves. It is used by ships to detect objects below the surface of the water.

A ship using sonar sends out a sound wave. Then it records the echo of the sound after it is reflected by an underwater object.

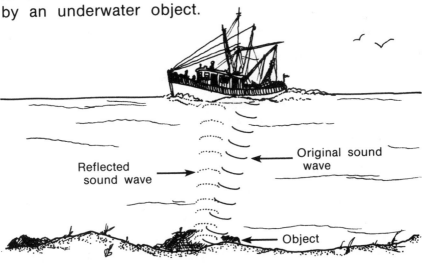

By measuring the time required for the echo to arrive back at the ship, scientists can calculate how far away the object is.

Recall the formula for calculating distance.

$$\text{DISTANCE} = \text{RATE} \times \text{TIME}$$
or
$$D = R \times T$$

Suppose we wanted to know the distance of an underwater object. Multiply the speed of sound in water by the time it takes for the echo to return. Then divide that answer in half to find the distance to the object. It is necessary to divide by 2 since the sound wave traveled *to* the object and *back*, while we want to know only the distance *to* the object. Study the example.

Example: How far away is an object whose echo takes 4 seconds to return? The speed of sound in water is 1497 meters per second.

Method: D = R × T
D = 1497 meters per second × 4 seconds
D = 5988 meters

Now divide by 2.

Distance = 5288 meters ÷ 2
Distance = 2644 meters

▶ *Practice.* Answer these questions about the distance of objects underwater.

1) How far away is an object whose echo takes 6 seconds to return?

2) How far away is an object whose echo takes 9 seconds to return?

3) How far away is an object whose echo takes 10 seconds to return?

4) How far away is an object whose echo takes 1 second to return?

INVESTIGATION 7

Purposes: To use sound waves to communicate, using a tin-can telephone; to show that sound is caused by vibration.

A. Materials:

Quantity:	Item Description:
2	Tin cans (or styrofoam cups)
2	Buttons
15 meters	String
1 stick	Paraffin wax (not essential)

B. Procedure:

1) Construct the tin-can telephone system as shown in the following diagram.

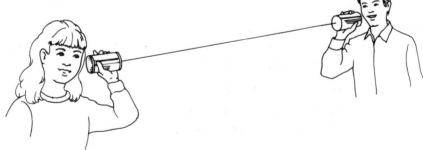

2) Punch a small hole in the bottom of a tin can. If using a styrofoam cup, punch a hole with a small nail.

3) Feed one end of the string through the hole and tie a button on the end of the string.

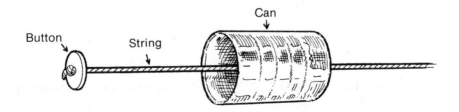

4) Do the same with both of the cans.

5) Pull the button into the bottom of the can.

6) Pull the string tight between the two cans.

7) While one person listens, the other person speaks into his or her tin can.

C. Interpretation and Further Discovery

1) While talking, squeeze the string with your fingertips. What happens to the sound at the other end? Why?

2) Will your telephone work around corners? Why or why not?

3) Use different materials in place of the string. Some other materials to try are thread or fishing line. Does this have any effect on the phone's operation? Try some large twine. What problems do you encounter?

4) Try the following "party line" telephone. Does it work?

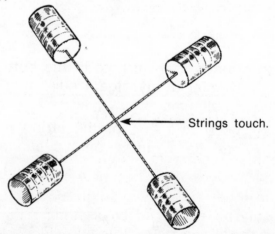

Strings touch.

5) Try waxing the string by rubbing it with the paraffin wax. This usually improves the sound quality of the phone. Why would this be?

10.8 LIGHT

We see objects because of the *LIGHT* that is reflected from them. Light is a form of energy that can be seen by our eyes. Most of this light is produced by the sun, a glowing ball of extremely hot gases. Light is produced by objects which are at high temperatures.

As an object becomes hotter, it begins to glow and radiate light. At first, the color of the object may be red. Then it changes to orange, and then yellow, as it becomes hotter. Finally, the object becomes white or blue-white. The hottest objects often have a blue-white color due to their extremely high temperatures.

Stars are a good example of objects that produce light. Light is radiated from stars because of their high temperatures. In fact, the color of a star and its temperature are related. This is shown in the chart:

Temperature and Color of Stars	
Surface Temperature (°C)	Color of Star
20,000	blue-white
9,000	white
6,000	yellow
3,000	red

10.9 THE NATURE OF LIGHT

Light is a complicated form of energy. Some experiments seem to suggest that light is made of tiny particles. These particles have been named *PHOTONS*. But, in other experiments, light shows properties which can only be explained by assuming that light consists of waves. Because of this two-sided nature, scientists describe light as having a dual nature. Sometimes light can be explained in terms of particles. At other times it can be explained in terms of waves. Most of the properties can be described in terms of the wave nature of light.

10.10 COLOR

Sir Isaac Newton was one of the first scientists to study the nature of light. By using a glass device in the shape of a long triangle, he was able to show that ordinary light was in fact a combination of all of the colors of the rainbow. This glass device is called a *PRISM*.

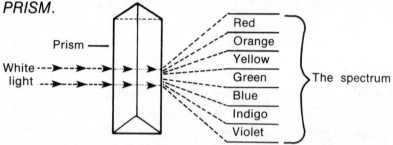

The combination of these colors of the rainbow is called the *SPECTRUM* of light. The order of the colors is important because they always appear in that order. A convenient way to remember the order is by memorizing this name of an imaginary person:

R O Y   G   B I V

The letters in the name each stand for a color: Red, Orange, Yellow, Green, Blue, Indigo, Violet.

These colors are the same ones you see when looking at a rainbow. In fact, the rainbow is caused by the tiny droplets of rain acting as prisms and creating the spectrum in the sky.

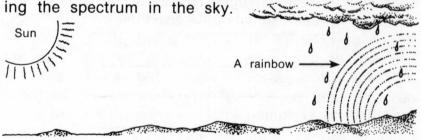

10.11 THE REFRACTION OF LIGHT

REFRACTION refers to a wave which is bent. For example, light is refracted when it travels from one substance into another. Light that enters glass after traveling through air is refracted as it passes into the glass.

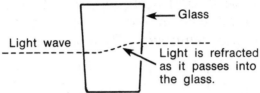

Glass

Light wave

Light is refracted as it passes into the glass.

You can also observe refraction by placing a straw in a glass of water. The straw will appear to be bent when observed from the side. The light rays from the straw are bent as they travel from the water into the air.

The amount of bending depends on the material through which the light is traveling. A number known as the *INDEX OF REFRACTION* tells how much the light is bent for various materials. The larger the number, the more light is bent by that substance. Here is a table listing the index of refraction for some common substances.

Index of Refraction for Various Substances	
Air	1.0
Glass	1.5
Water	1.3

As light passes through materials, each of the different colors is refracted by a different amount. It is because of this difference that light is spread out into a spectrum by a prism.

10.12 LIGHT CAN BE REFLECTED BY MIRRORS

Just as sound can be reflected, light also can be reflected from certain objects. One common object which reflects light is a *MIRROR*. One type of mirror is a *PLANE MIRROR*. A plane mirror is one whose surface is flat and smooth. A common pocket mirror is an example of a plane mirror.

When light strikes a plane mirror, it is reflected back by the silvered surface of the mirror. The rays of light which strike the mirror are called *INCIDENT* rays. The rays which bounce off the mirror are called *REFLECTED* rays.

The angle at which the incident rays strike the mirror is always equal to the angle of the reflected rays. These angles are measured from a line which is drawn perpendicular from the surface of the mirror, as in the following diagram.

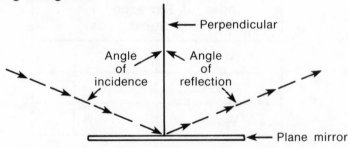

Images

When you look at your reflection in a plane mirror, what you see is called an *IMAGE*. Take a minute and observe yourself in a plane mirror. Move your right arm and notice which arm of the image moves. It is the left arm of the image! The image has been reversed from right to left.

Image in a plane mirror is reversed from right to left.

The image in a plane mirror is called a *VIRTUAL* image. Virtual means "not real." A virtual image appears to be behind the mirror. A virtual image cannot be projected onto a screen.

The other type of image is called a *REAL* image. It can be projected onto a screen. If you have a special rounded mirror called a shaving mirror, you will find that you can project images on a wall with one of these mirrors.

Types of Mirrors

Concave Mirror. We have already mentioned the flat, plane mirror. The shaving mirror is an example of a type known as a *CONCAVE* mirror. A concave mirror "caves" in on the mirror side.

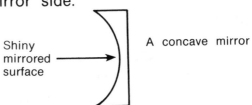

Shiny mirrored surface

A concave mirror

A concave mirror will *FOCUS* light, which means that the light rays come together at one point. To focus means "to make clear." The word *focus* can also be a noun. The focus is the point where light rays come together.

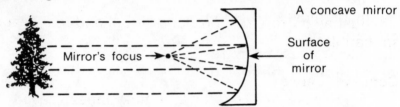

A concave mirror

Mirror's focus →

Surface
of
mirror

In a concave mirror, if your eye is between the focus of the mirror and the surface of the mirror, you will see a virtual image which is *MAGNIFIED* or enlarged. If your eye is positioned further away from the mirror than the focus point, you will see a real, upside down image.

Convex Mirror. Another type of mirror is called a *CONVEX* mirror. In a convex mirror, the mirror side is curved outward.

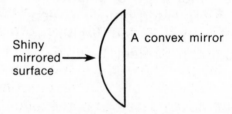

Shiny
mirrored →
surface

A convex mirror

A convex mirror creates a virtual image which looks smaller than the real object. However, you can see much more area in a convex mirror. This type of mirror is often used as a rear-view mirror on automobiles.

10.13 LENSES

A *LENS* is a piece of glass or plastic that is used in optical instruments. Light passes through a lens. Light, while reflected in mirrors, is refracted, or bent, in lenses.

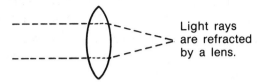

Light rays
are refracted
by a lens.

Just as mirrors can produce virtual and real images, so can lenses.

Concave Lenses

A concave lens has one or both sides "caved" in.

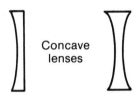

Concave
lenses

Concave lenses will spread light waves out. The image formed by a concave lens is a virtual image. It cannot be projected onto a screen. Concave lenses are used often in the making of glasses for people who are *NEARSIGHTED*. Nearsighted means that you can't see far away objects clearly. In a nearsighted person's eye, the natural lens of the eye forms an image in front of the retina instead of on it. This produces a fuzzy image. By using glasses made from a concave lens, the light is refracted out before entering the eye. When the light now passes into the eye, a proper image is formed on the retina.

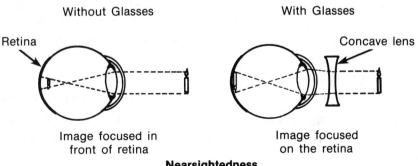

Without Glasses

With Glasses

Retina

Concave lens

Image focused in
front of retina

Image focused
on the retina

Nearsightedness

Convex Lenses

Convex lenses are lenses which are thicker in the middle than they are on the ends.

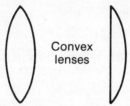

Convex lenses refract light, also. However, in a convex lens the light is refracted to a focus.

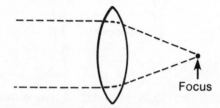

The convex lens can be used to magnify images by holding the lens close to the eye and observing an object. The enlarged image is a virtual image which is larger than the object.

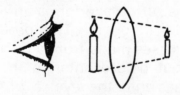

Another type of image, the real image, can be formed by a convex lens, also. Hold a convex lens far from your eye and observe an object at a distance. The image you now see is upside down and real. It can be projected onto a screen.

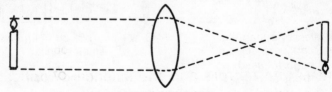

Convex lenses are used in making eyeglasses, just as concave lenses are. Convex lenses are used to correct the vision of farsighted people. A farsighted person cannot clearly see objects which are near to him or her. In the farsighted eye, the image is fuzzy because the retina is too close to the eye's lens. There isn't enough room for the image to be focused properly. By using the convex lens in front of the eye, the image can be focused properly on the retina.

No Glasses   With Glasses

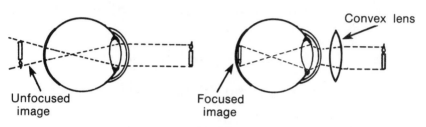

Convex lens

Unfocused image   Focused image

Farsightedness

10.14 PUTTING LENSES AND MIRRORS TO WORK

Many instruments make use of lenses and mirrors. They are called *OPTICAL* instruments because *optics* is the science of light. Telescopes and microscopes are used to enlarge images so that they can be studied in more detail.

A telescope   A microscope

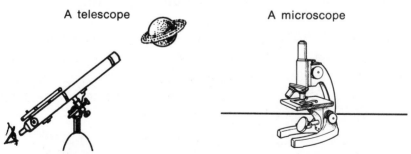

A telescope magnifies astronomical objects.

A microscope magnifies small objects.

Lenses and mirrors and the science of optics have greatly enlarged our knowledge of the world.

SUMMARY OF CHAPTER 10

1) Acoustics is the science of sound.

2) Sound is caused by vibrations.

3) Vibrating objects create sound waves.

4) Sound travels at different speeds through different materials.

5) The loudness of a sound is measured in decibels.

6) Frequency measures the number of vibrations per second.

7) A cycle is one complete vibration.

8) Pitch describes the highness or lowness of a sound.

9) An echo is a reflected sound.

10) Sonar is used by ships to measure the distance of objects under water.

11) Light has both wave properties and particle properties.

12) Particles of light are called photons.

13) White light can be broken into a spectrum by a prism.

14) The spectrum colors are red, orange, yellow, green, blue, indigo, and violet.

15) A rainbow is a spectrum caused by rain drops.

16) Refraction is the bending of light waves.

17) The index of refraction measures how much light is bent by a substance.

18) A mirror is an object which reflects light.

19) Three types of mirrors are the plane, convex, and concave mirror.

20) Two types of lenses are convex and concave.

21) An image is either real or virtual.

22) Lenses can be used to correct vision problems.

23) Telescopes and microscopes are instruments that use mirrors and lenses.

24) The science of light is called optics.

Vocabulary Words

acoustics	mirror
vibrate	incident
sound wave	index of refraction
decibel	plane mirror
frequency	convex
cycle	image
pitch	virtual image
reflection	concave
echo	nearsighted
sonar	focus
light	farsighted
optics	magnify
photon	real image
prism	microscope
spectrum	telescope
refraction	

• CHAPTER REVIEW EXERCISES •

▶ Copy the sentences. Fill in the missing words.

1) An object that breaks light into a spectrum is called a _____.

2) The number of cycles per second is called _____.

3) An instrument which magnifies astronomical objects is called a _____.

4) A flat mirror is called a _____ mirror.

5) A reflected sound wave is an _____.

▶ Write the letter of the best answer for each item.

6) Particles of light are called:
a) protons.　　b) photons.　　c) electrons.

7) The intensity of a sound is measured in:
a) decibels.　　b) cycles.　　c) vibrations.

8) Sound is caused by:
a) diffraction.　　b) echoes.　　c) vibrations.

9) The method of measuring distances under water is known as:
a) convex.　　b) sonar.　　c) refraction.

10) Which of the following is a type of image?
a) convex　　b) real　　c) reflection

Calculator Practice

▶ Use your calculator to determine the distance of each of these objects under water. The time it takes for the echo to return is given. The speed of sound in water is 1497 meters per second. Remember to divide by 2 for the final answer.

	Time in Seconds	Distance		Time in Seconds	Distance
1)	4.2	?	2)	3.89	?
3)	18	?	4)	12.63	?
5)	.03	?	6)	.86	?
7)	2.3	?	8)	8.36	?

Computer Program

▶ This program calculates the distance of objects under water.

```
10 CLS
20 PRINT "THIS PROGRAM CALCULATES THE"
30 PRINT "DISTANCE OF UNDERWATER OBJECTS."
40 PRINT "PLEASE NOTE THAT THESE ANSWERS"
50 PRINT "ARE ONLY APPROPRIATE FOR A WATER"
60 PRINT "TEMPERATURE OF 25 DEGREES CELSIUS."
70 PRINT:PRINT
80 PRINT "HOW MANY SECONDS DID IT TAKE"
90 PRINT "FOR THE ECHO TO RETURN?";
100 INPUT T
110 LET D=(1497 * T)/2
120 PRINT:PRINT
130 PRINT "THE OBJECT IS ";D;" METERS AWAY."
140 PRINT:PRINT
150 PRINT "DO YOU WANT TO DO"
160 PRINT "ANOTHER PROBLEM?(Y/N)"
170 INPUT R$
180 IF R$="Y" THEN 10
190 CLS
200 END
```

Chapter 11

Electricity

Chapter Goals:

1. To list three conductors and three insulators of electricity.

2. To identify open and closed circuits.

3. To identify various kinds of batteries.

4. To describe the difference between batteries in parallel and batteries in series.

5. To describe characteristics of series and parallel circuits.

6. To use Ohm's Law in problem solving.

7. To list two devices that measure electricity.

8. To describe how electricity can be controlled.

9. To calculate the power used in a circuit.

10. To describe the difference between A.C. and D.C. electricity.

11.1 STATIC ELECTRICITY

Most of us are familiar with static electricity. Static electricity causes the clothes from the dryer to cling together. Have you ever gotten a shock when you touched metal after you walked across a carpet? That was caused by static electricity. Your walking caused heat and friction on the carpet, and this made electrons in the carpet more active. Electrons left the carpet and entered your body. When your charged body touched the metal, the extra electrons jumped from your finger. When electrons move from one place to another, energy is transferred.

11.2 ELECTRIC CURRENT

The movement of electrons from one place to another is an electric *CURRENT*. Currents from static electricity are not easy to control. An electrical current produced by a battery, however, is predictable and easy to use.

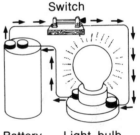

Switch

In the diagram at the right, electric current flows from the battery. It runs along a metal wire, through a switch, lights up a bulb, and returns to the battery.

Battery   Light bulb

Conductors and Insulators

The electric current in the example above flows through a metal wire. Any material that allows electrons to flow through it easily is a *CONDUCTOR*. Metals are good conductors. Poor conductors include glass, rubber, and plastic. Because these items do not conduct electricity well, they are good *INSULATORS*.

Good Conductors	Good Insulators
Copper wire Aluminum Gold Silver Water with minerals	Glass Plastic Rubber Dried wood

Your living room table lamp uses copper wire as a conductor for the electricity. A rubber or plastic coating covers the copper wire to prevent fire or accidental electric shock. The covering is called the *INSULATOR*.

Insulator

Conductor

11.3 ELECTRIC CIRCUITS

An electric current that is generated, or produced, by a battery moves in only one direction. Therefore, it is known as *DIRECT CURRENT*. "Direct current" is often shortened to "D.C." Study the diagrams below that show the paths of direct current.

The path that the current follows is called a *CIRCUIT*. The current begins at the battery, goes through the circuit, and returns to the battery.

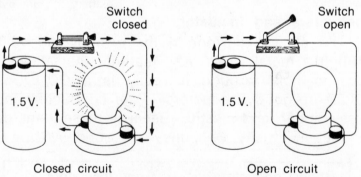

Switch closed	Switch open
1.5 V.	1.5 V.
Closed circuit	Open circuit

In the first diagram the switch is closed, allowing the current to pass through. This is a *CLOSED* circuit.

In the second diagram the switch is open. The current cannot pass through it, so the bulb does not light. This is an *OPEN* circuit.

11.4 BATTERIES

Electrical circuits need a source of electricity. This source can be either a *BATTERY* or electricity generated by an electric company. A battery is a device that changes chemical energy into electrical energy. There are two main types of batteries: dry and wet. Dry batteries are the type that are used in flashlights, radios, and other small appliances. Wet batteries are used in cars and motorcycles.

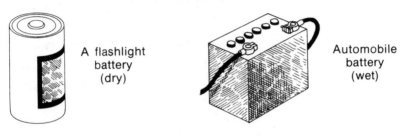

A flashlight battery (dry)

Automobile battery (wet)

Dry Batteries

Dry batteries come in many sizes and shapes. The most common is known as a D size battery. It is used in flashlights and large radios. Other sizes are C, AA (called double A), and AAA (called triple A) batteries. Generally, the smaller batteries are used with devices that don't require much power.

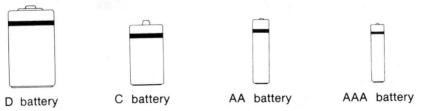

D battery   C battery   AA battery   AAA battery

Regardless of the size of a dry battery, each of them is made of a zinc container filled with a black moist powder known as manganese dioxide. In the center of the battery is a long rod made of the element carbon. The top of the battery is sealed with a special type of epoxy to keep the battery from leaking.

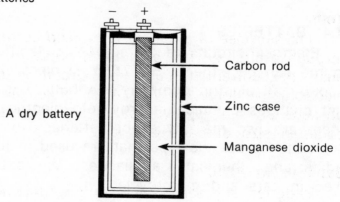

A dry battery

Carbon rod

Zinc case

Manganese dioxide

Each battery has two *TERMINALS*, or points, where they are connected to an appliance. Larger, hobby-type batteries have both terminals on the top. The center terminal is the *POSITIVE* terminal. It will be marked with a "+." The other terminal is the *NEGATIVE* terminal. This is labeled "−."

Smaller batteries have only one button on top. This is the positive terminal. The bottom metal part of the battery is the negative terminal. Electrons move from the negative terminal to the positive terminal.

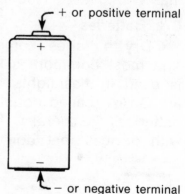

+ or positive terminal

+

−

− or negative terminal

Wet Batteries

Wet batteries are different from dry batteries because they are filled with a liquid, rather than a dry chemical. Most wet batteries are made from a hard rubber case which is filled with a solution of sulfuric acid. Plates, made of lead, are placed inside of the liquid. The terminals on a wet battery are clearly marked as "+" and "−" or, sometimes, as POS (for positive) and NEG (for negative).

Volts

The purpose of a battery is to keep current moving through a circuit. The amount of pressure needed to push the current through the circuit is called the electromotive force, or *VOLTAGE*. The circuit shown on page 212 has a 1.5 volt battery. 1.5 volt is the most common size of battery.

Batteries in Parallel

Batteries are sometimes connected in *PARALLEL* to do the work of a larger battery. The wire connects the terminals that are alike from one battery to another. This forms a parallel connection. A parallel connection gives the electrical current additional capacity. It does not increase the voltage, however.

Look at the diagrams below. The second circuit has two batteries connected in parallel. The bulb in this circuit will stay lit longer because the capacity of this circuit is greater than that of the first circuit.

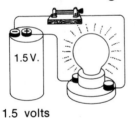

1.5 volts

Total voltage is 1.5 volts.

The second circuit has two 1.5 batteries. The total voltage is still 1.5 volts. Batteries connected in parallel allow the voltage to remain the same, while increasing the capacity.

These batteries are connected in parallel.

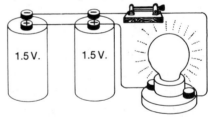

The total voltage is 1.5 volts.

Batteries in Series

Another way of connecting batteries is in *SERIES*. This is done by connecting the positive terminal of one battery with the negative terminal of another battery. Batteries in series increase the voltage of the circuit. To calculate the voltage of batteries in series, add the voltages of the batteries together.

In the circuit shown below, the batteries are in series. A short jumper wire connects a positive and a negative terminal. A second wire connects the lamp and switch to the batteries.

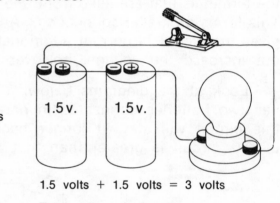

Batteries in series

1.5 volts + 1.5 volts = 3 volts

Total voltage is 3 volts.

Flashlight batteries are in series. The positive terminal of one battery touches the negative terminal of the other battery.

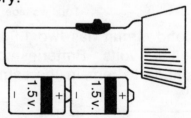

1.5 volts + 1.5 volts = 3 volts

The common 9-volt transistor radio battery is made by placing six small 1.5-volt batteries in series. 6×1.5 volts = 9 volts.

▶ **Practice.**   Answer these questions.

1) What is a current?

2) What is an insulator?

3) Name three conductors.

4) What does D.C. mean?

5) Name three insulators.

6) When two batteries with the same voltage are connected in parallel, what happens to the voltage?

7) When two batteries are connected in series, what happens to the voltage?

8) Which of the following circuits is a closed circuit? Which is an open circuit?

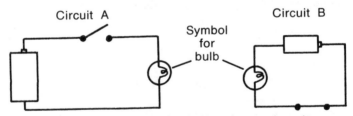

9) Identify each of the following circuits as either a parallel or series connection of batteries.

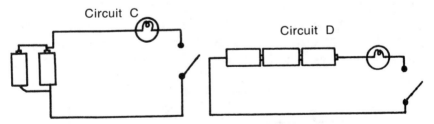

10) What are four sizes of dry batteries?

11) Where is the positive terminal of a dry battery usually located?

11.5 SERIES CIRCUITS

Just as batteries can be connected in series, appliances also can be connected this way. For example, decorative tree lights are all connected together with one common plug. Many of these strings of lights are connected in *SERIES*. In a series circuit, the electrons must pass through each light bulb. Study the following diagram.

A series circuit

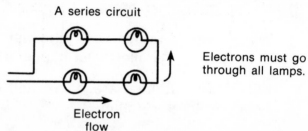

Electrons must go through all lamps.

Electron flow

An interesting property of a series circuit is that if one light is unscrewed or blows out, ALL of the lights will go out. If a light bulb is unscrewed or blows out, the circuit is opened, and electrons cannot flow.

An open series circuit

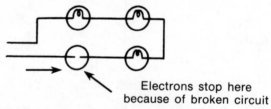

Electrons stop here because of broken circuit

A series circuit can be identified by tracing the path of the circuit. If there is only ONE path around the circuit, then it is a series circuit.

Series circuits

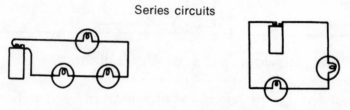

There is only one path around the circuit!

Characteristics of Series Circuits

If one bulb is connected to a battery, the bulb will shine brightly. If another bulb is added in series with the circuit, each of the bulbs will light, but they will each be much dimmer. You can continue adding bulbs in series, but as each one is added, the brightness of each bulb will be less. Eventually, all the bulbs will be too dim to see.

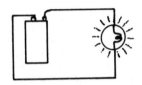

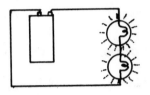

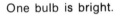

One bulb is bright.   Two bulbs are dimmer.

A Practical Use for a Series Circuit

Most decorative tree lights are no longer wired in series because it is too difficult to find out which light bulb is bad when the bulbs go out. Each bulb must be tested individually to determine the one that is bad.

However, there are other practical uses for series circuits. For example, your house is protected by *FUSES* or *CIRCUIT BREAKERS* which are part of a series circuit. The purpose of a fuse or circuit breaker is to protect your house from a fire. Look at the following diagram of a light and a fuse together in a series circuit.

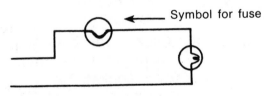

A fuse in a series circuit

The fuse is made from a piece of metal designed to melt at a certain temperature. As the current in a circuit increases, the wires begin to heat up. If they get too hot, the wires could start a fire in your home! Before this happens, however, the fuse will melt and break the circuit. Thus a fire is prevented. When a fuse melts, it cannot be reused. A circuit breaker, on the other hand, can be reset and used again after the faulty circuit is repaired.

▶ *Practice.* Answer these questions.

1) How can a series circuit be recognized?

2) What is a fuse?

3) What is the difference between a fuse and a circuit breaker?

4) What happens to the brightness of bulbs when more than one is included in a series circuit?

INVESTIGATION 8

Purpose: To list characteristics of a series circuit.

A. Materials:

Quantity:	Item Description:
2	D batteries
2	Holders for batteries
3	Flashlight type bulbs
3	Bulb sockets
5 feet	Common bell wire
1	Switch (any type)

B. Procedure:

1) Make each of the circuits shown in A, B, and C.

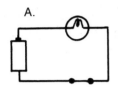

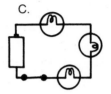

2) Answer questions 1, 2, and 3 in the interpreta-
tion as you complete the circuits.

3) Make each of the circuits shown in D and E.

4) Answer questions 4, 5, and 6 in the interpreta-
tion.

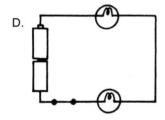

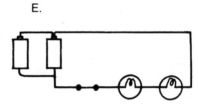

C. Interpretation:

1) What happens in circuit B when one of the
lights is unscrewed?

2) What happens in circuit C when one of the
lights is unscrewed?

3) What happens to the brightness of the bulbs
in circuits B and C compared to circuit A?

4) Are the batteries in diagram D connected in
series or parallel?

5) Are the batteries in diagram E connected in
series or parallel?

11.6 PARALLEL CIRCUITS

The lights and appliances in your home are not wired together in a series circuit. If they were, every time a bulb burned out, all the other lights and appliances would not work! Instead, most of the circuits in houses are *PARALLEL* circuits. Look at the following diagram of two lamps connected in parallel.

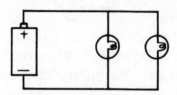

You can identify a parallel circuit because there is more than one path around the circuit.

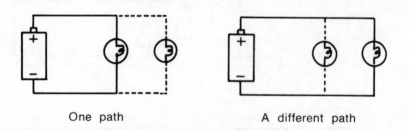

One path   A different path

As you can see in the above diagram, there are two paths around this circuit. It is therefore a parallel circuit.

Characteristics of Parallel Circuits

Because there is more than one path for the electrons to flow in a parallel circuit, when one bulb burns out, the other bulbs will stay lit.

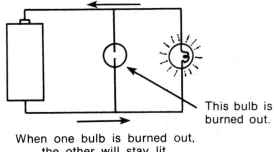

This bulb is
burned out.

When one bulb is burned out,
the other will stay lit.

When several bulbs are connected in parallel, all the bulbs will remain as bright as just one bulb would in a circuit. Even though there are more paths for the electric current to flow through, all the bulbs still remain bright. However, because of the increased current being used, the battery will not last as long as in a series circuit.

▶ *Practice.* Answer these questions.

1) How can you recognize a parallel circuit?

2) What happens to the bulbs in a parallel circuit when one bulb burns out?

3) Determine the number of paths in each of the following parallel circuits.

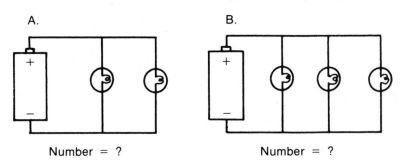

A.

Number = ?

B.

Number = ?

4) What happens to the brightness of bulbs in a parallel circuit when more bulbs are included?

5) Identify each of the following circuits as either parallel or series.

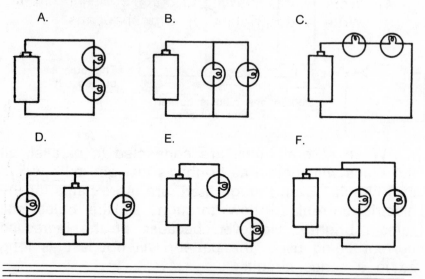

A.   B.   C.

D.   E.   F.

INVESTIGATION 9

Purposes: To observe the characteristics of various circuits. To predict which circuits are closed.

A. Materials:

Quantity:	Item Description:
2	D size batteries
2	Flashlight bulbs
5 feet	Common bell wire, cut into needed lengths

B. Procedure:

1) Draw each of the following circuits on your lab paper.

2) Construct each circuit. Use the materials listed above.

3) Predict whether or not you think the bulb or bulbs will light.

4) Work with a partner to complete each circuit. Write a description of what happens.

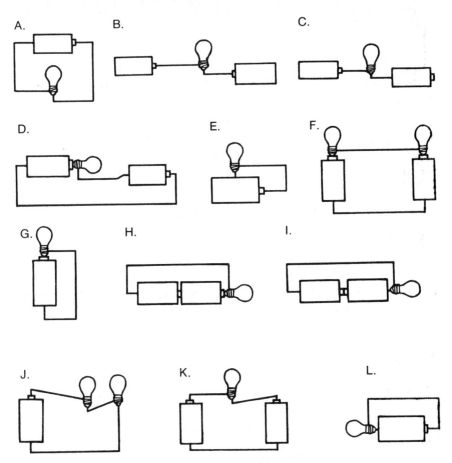

11.7 THE COMPONENTS OF ELECTRICITY

Current

The flow of electrons through a conductor is called the current. The unit of measure for current is the *AMPERE*. The amount of current an electric iron uses is about ten amperes. The word *amperes* can be shortened to "amps."

Potential Energy

How much force does an electric current have? The electromotive force, or push, is measured in *VOLTS*. A circuit with 50 volts has twice the force of a circuit with 25 volts.

Resistance

Anything that prevents the electrons from moving along quickly in an electric current is called *RESIST-ANCE*. A light bulb offers resistance, for example. So does some material. When material, such as a wire, has very little resistance, it is a good conductor. Resistance is measured in *OHMS*. As the resistance increases, the amperes decrease. The ohm was named for the German scientist, Georg Ohm. The abbreviation for ohm is the Greek letter omega (Ω).

11.8 OHM'S LAW

The current, the potential energy, and the resistance of a circuit are all related. Georg Ohm made a mathematical formula that described this relationship. We call his formula "Ohm's Law." The three simple formulas below show how Ohm's law works.

Current = Voltage ÷ Resistance

Resistance = Voltage ÷ Current

Voltage = Current × Resistance

Scientists use letters to stand for these words in formulas:

Current — I

Resistance — R

Voltage — E

11.9 AMPERES

The diagram below shows an electric circuit. It has a 1.5 volt battery and a lamp. A lamp is an example of a *RESISTOR*, a device that offers resistance to the current.

Example 1:

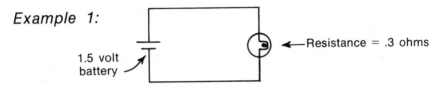

1.5 volt battery

Resistance = .3 ohms

How much current is in this circuit?

Current = Voltage ÷ Resistance

$$Current = \frac{1.5 \ volts}{.3 \ ohms}$$

Current = 5 amperes

Here is a simple figure to help you remember how to work with Ohm's law:

When you need to find the current, place your finger over the I. You will see the fraction $\frac{E}{R}$. This indicates "voltage divided by resistance."

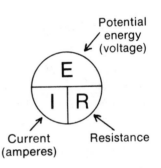

Potential energy (voltage)

Current (amperes)   Resistance

Example 2:

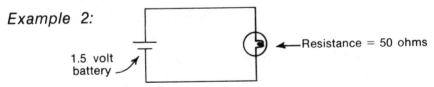

1.5 volt battery

Resistance = 50 ohms

What is the current?

$$I = \frac{E}{R} = \frac{1.5 \ volts}{50 \ ohms} = .03 \ amperes$$

Example 3: Connect two flashlight batteries in series to get 3 volts for this circuit:

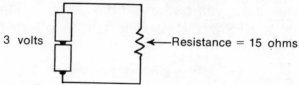

3 volts —Resistance = 15 ohms

What is the current?

$$I = \frac{E}{R} = \frac{3 \text{ volts}}{15 \text{ ohms}} = .2 \text{ amperes}$$

▶**Practice.** Find the current for each problem below. The voltage and the resistance are given. Round the amperes to the nearest tenth.

1) E = 10 volts
   R = 5 ohms
   I = ?

2) E = 15 volts
   R = 5 ohms
   I = ?

3) E = 50 volts
   R = 5 ohms
   I = ?

4) E = 35 volts
   R = 10 ohms
   I = ?

5) E = 25 volts
   R = 10 ohms
   I = ?

6) E = 100 volts
   R = 12 ohms
   I = ?

11.10 RESISTANCE

The diagram below shows an electric circuit that uses 120 volts. It has a current of 10 amperes. A toaster is plugged in. How much resistance does it offer?

Example 1:

I = 10 amperes

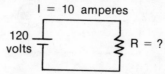

120 volts   R = ?

Resistance = Voltage ÷ Current

$$\text{Resistance} = \frac{120 \text{ volts}}{10 \text{ amperes}}$$

Resistance = 12 ohms

When you need to find the resistance, you may use the circle chart. Place your finger over the R. You will see the fraction $\frac{E}{I}$. This indicates "voltage divided by current."

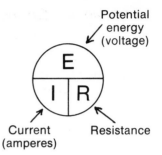

Potential energy (voltage)

Current (amperes)

Resistance

Example 2: What is the resistance?

I = 3.5 amperes

E = 40 volts

R = ?

$$R = \frac{E}{I}$$

$$R = \frac{40 \text{ volts}}{3.5 \text{ amperes}} = 11.4 \text{ ohms}$$

▶**Practice.** Find the resistance for each problem below. The voltage and the current are given. Round your answer in ohms to the nearest tenth.

1) E = 50 volts
    I = 5 amperes
    R = ?

2) E = 25 volts
    I = 5 amperes
    R = ?

3) E = 70 volts
    I = 10 amperes
    R = ?

4) E = 1.5 volts
    I = 5 amperes
    R = ?

5) E = 3 volts
    I = .5 ampere
    R = ?

6) E = 1.5 volts
    I = .2 ampere
    R = ?

7) E = 20 volts
    I = 1.5 amperes
    R = ?

8) E = 30 volts
    I = 4 amperes
    R = ?

9) E = 1.5 volts
    I = .3 ampere
    R = ?

10) E = 3 volts
    I = .1 ampere
    R = ?

11.11 VOLTAGE

The diagram below shows an electric circuit that has a current of 2 amperes. The resistance is 5 ohms. How much voltage is in the circuit?

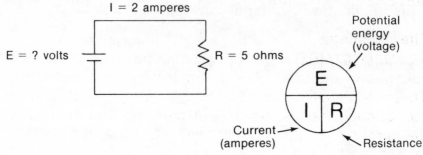

I = 2 amperes

E = ? volts

R = 5 ohms

Potential energy (voltage)

E

I | R

Current → (amperes)

Resistance

You may use the circle chart. When you need to find voltage, place your finger over the E. You will see I|R. This indicates "current times resistance."

$$\text{Voltage} = \text{Current} \times \text{Resistance}$$

$$E = I \times R$$

$$E = 2 \text{ amperes} \times 5 \text{ ohms, or } 10 \text{ volts}$$

$$E = 10 \text{ volts}$$

▶ *Practice.* Find the voltage of each problem below. The current and the resistance are given.

1) I = 3 amperes
   R = 6 ohms
   E = ? volts

2) I = 6 amperes
   R = .5 ohm
   E = ? volts

3) I = .3 ampere
   R = .7 ohm
   E = ? volts

4) I = 6 amperes
   R = .1 ohm
   E = ? volts

5) I = 1 ampere
   R = 10 ohms
   E = ? volts

6) I = .2 ampere
   R = 5 ohms
   E = ? volts

11.12 MEASURING ELECTRICITY

People who work with electricity often need to measure certain properties of circuits to determine if they are working properly. These people use *METERS* to measure the electricity in circuits.

The Voltmeter

A *VOLTMETER* is an instrument that measures voltages in circuits. The voltmeter has two terminals. These terminals must be connected in parallel with the circuit that is being measured. The following diagram shows the correct way of using a voltmeter in a circuit.

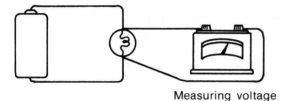

The voltmeter is connected in parallel with the bulb.

Measuring voltage

Voltmeters come in many different sizes. Some measure high voltages. Some can only measure low voltages.

The Ammeter

An *AMMETER* is used to measure the amount of current flowing in a circuit. Unlike the voltmeter, the ammeter must be connected in series with the circuit being measured. It also has two terminals with which it is connected to the circuit.

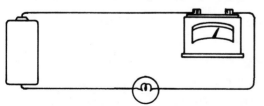

The ammeter is connected in series with the bulb.

A special kind of meter that can measure very small currents is called a *GALVANOMETER*.

11.13 CONTROLLING ELECTRICITY

Electricity is much more useful to us than other forms of energy because it is easy to control. It can be turned on and off by placing switches in the circuits. Using switches allows us to use the electricity only when we need it.

Types of Switches

There are many different types of switches available for controlling circuits.   The easiest type to use for experiments is called a *KNIFE* switch.

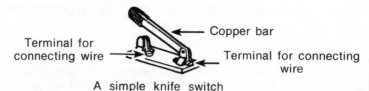

Terminal for connecting wire ← Copper bar

Terminal for connecting wire

A simple knife switch

There are many kinds of switches. The purpose of a switch is to turn an appliance on or off by opening or closing the circuit.

A doorbell switch   A wall switch   A lamp switch

Controlling Current in a Circuit

Do you have a special type of controller in your house known as a light dimmer? This device allows you to change or vary the amount of resistance in a circuit.   By changing the resistance, the brightness of the bulb is changed.

Inside the light dimmer is a special type of high resistance wire connected to a sliding piece of metal. As the slider moves over the wire, the resistance is changed. Remember that changing the resistance in a circuit changes the amount of current flowing in the circuit.

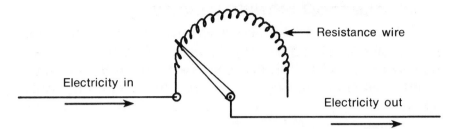

A light dimmer set to low resistance

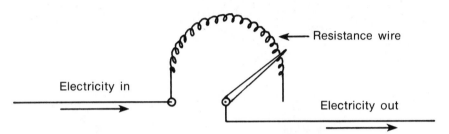

A light dimmer set to high resistance

Another name for a light dimmer is a *RHEOSTAT*. Some other examples of uses for rheostats are as volume controls on radios and TVs, heat controls on an oven, and the controller on an electric train transformer.

► *Practice.*   Write the answers.

1) What do we use to measure the electricity in a circuit?

2) What does an ammeter measure?

3) How is a voltmeter connected in a circuit?

4) What is a galvanometer?

5) How is an ammeter connected in a circuit?

6) Name three types of switches.

7) What is a rheostat?

8) Name two appliances that use rheostats.

11.14 ELECTRIC POWER

Have you ever had to buy light bulbs at the store? If you have, you may have noticed that bulbs have a number stamped on the top, such as 60W or 100W. The W stands for *WATTS*. Watts are the units which measure the amount of electricity used. In other words, watts measure *POWER*.

A 100-watt light bulb uses more electricity than a 60-watt bulb. Generally, a high wattage bulb is brighter than one of lower wattage.

100-watt bulb

60-watt bulb

To find the power used in a circuit, you must first know the voltage and the amount of current (amperes) flowing in the circuit. These can be measured using electric meters. Once these values are known, the power can be calculated by using the following formula:

$$P = E \times I$$

POWER = VOLTAGE × CURRENT (AMPERES)

Study the following examples for finding power.

1) How much power is used by a light bulb if the voltage is 120 volts, and a current of .5 ampere is flowing in the circuit?

    Solution: $P = E \times I$

                  $P = 120$ volts × .5 ampere

                  $P = 60$ watts

2) How much power is used by an iron if the voltage is 120 volts, and a current of 10 amperes is flowing in the circuit?

> *Solution:* P = E × I
>
> P = 120 volts × 10 amperes
>
> P = 1200 watts

▶ **Practice.** Find the power in each of these circuits. The voltage and the current are given.

1) Voltage = 120 volts
   Current = 5 amperes
   Power = ?

2) Voltage = 115 volts
   Current = .5 ampere
   Power = ?

3) Voltage = 110 volts
   Current = .2 ampere
   Power = ?

4) Voltage = 110 volts
   Current = .1 ampere
   Power = ?

5) Voltage = 240 volts
   Current = 10 amperes
   Power = ?

6) Voltage = 208 volts
   Current = 20 amperes
   Power = ?

7) Voltage = 220 volts
   Current = 25 amperes
   Power = ?

8) Voltage = 220 volts
   Current = 50 amperes
   Power = ?

11.15 ELECTRICITY IN THE HOME

The electricity used in homes is not the same as that obtained from batteries. Batteries provide electricity known as *DIRECT CURRENT*, or D.C. for short. Direct current electricity flows in one direction through a conductor.

Direct current flows in one direction
through a conductor.

Direct current electricity is a convenient form of electricity to use for small appliances. It is not satisfactory for many other applications, however, especially where very high voltages are needed.

One other problem with D.C. electricity is that it is difficult to transmit it over long distances. Most power companies are located far away from residential areas. They must use long stretches of wire to get the electricity to our homes. To transmit electricity over such long distances, it is necessary to have high voltages.

The type of electricity in your home is known as *ALTERNATING CURRENT*, or A.C. for short. The electricity in A.C. circuits alternates, or switches, its direction of movement back and forth.

Alternating current flows back and forth in a conductor

In the United States, the A.C. current generated flows back and forth 60 times each second. This is known as 60-cycle A.C. current. In most European countries, the A.C. is supplied at 50 cycles. For this reason, most appliances made for use in the U.S. will not work properly in European countries.

Alternating current electricity can be sent over long distances much easier than direct current. The reason for this is that the low voltage can be changed easily to higher voltages.

11.16 TRANSFORMERS

A.C. electricity is much easier to use because its voltage can be changed. The voltage can be *IN-CREASED* using a *TRANSFORMER*. A transformer is made of an iron core which has turns of wire wrapped around it.

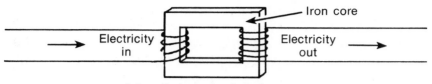

Internal view of a transformer

Notice that the wires coming into a transformer are not even connected to the wires coming out. The electric energy is transferred through the iron transformer without a direct connection!

A transformer is useful because it can change the voltage in a circuit. In the following diagram, the *PRIMARY* winding of the transformer is on the side where the electricity enters the transformer. In the diagram there are 5 turns of wire on this side. On the other side, called the *SECONDARY* winding, there are 10 turns of wire. The ratio of secondary turns to primary turns, 10 to 5, is 2. This transformer will multiply the incoming voltage by 2.

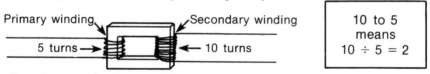

10 to 5 means 10 ÷ 5 = 2

Transformer showing primary and secondary windings

If the incoming voltage were 10 volts, then the outgoing voltage would be 20 volts (2 × 10 volts).

Study the following examples.

1) A transformer has a primary voltage of 12 volts. The primary winding has 100 turns of wire. The secondary winding has 200 turns. What is the secondary voltage?

   Solution: Find the secondary to primary ratio. The ratio is 200 to 100, or 2

   Secondary voltage = Primary volts × Ratio
   Secondary voltage = 12 × 2 = 24 volts

2) A transformer has a primary voltage of 10 volts. The primary winding has 500 turns of wire. The secondary winding has 1500 turns of wire. What is the secondary voltage?

Solution: Find the secondary to primary ratio.

The ratio is 1500 to 500. $\frac{1500}{500} = 3$

Secondary voltage = Primary volts × Ratio

Secondary voltage = 10 × 3 = 30 volts

Sometimes transformers are used to *DECREASE* the voltage, as in the following example.

3) A transformer has a primary voltage of 100 volts. The primary winding has 500 turns of wire. The secondary winding has 250 turns of wire. What is the secondary voltage?

Solution: Find the secondary to primary ratio.

The ratio is 250 to 500. $\frac{250}{500} = \frac{1}{2} = .5$

Secondary voltage = Primary volts × Ratio

Secondary voltage = 100 × .5 = 50 volts

▶ *Practice.* Solve these problems.

1) A transformer has a primary voltage of 20 volts. The primary winding has 200 turns of wire. The secondary winding has 400 turns of wire. What is the secondary voltage?

2) A transformer has a primary voltage of 30 volts. The primary winding has 750 turns of wire. The secondary winding has 250 turns of wire. What is the secondary voltage?

SUMMARY OF CHAPTER 11

1) Static electricity cannot be controlled easily.
2) An electric current is the movement of electrons from one place to another.
3) Conductors are materials that allow electricity to flow through them easily.
4) Insulators are materials that do not allow electricity to flow through them easily.
5) Electric current that flows in only one direction is called direct current, or D.C.
6) A circuit is the path through which a current flows.
7) A battery changes chemical energy into electrical energy.
8) The four common sizes of batteries are D, C, AA, and AAA.
9) Batteries have two terminals — positive and negative.
10) The electromotive force is measured in volts.
11) Batteries connected in parallel have more capacity than one battery. The voltage is not changed.
12) When batteries are connected in series, the voltages are added.
13) An electric circuit that has only one path is called a series circuit.
14) An electric circuit that has more than one path is called a parallel circuit.
15) A fuse or circuit breaker will protect a house from a fire.
16) The unit for electric current is the ampere.
17) The unit of resistance is the ohm.
18) The voltage, current, and resistance in a circuit are related to each other by Ohm's Law. $I = E/R$
19) A voltmeter measures the voltage of an electric circuit.

20) An ammeter measures the current flowing in an electric circuit.

21) A galvanometer measures very low currents.

22) Electricity can be controlled in a circuit by using a switch.

23) Electric power is measured in watts.

24) Electricity in the home is alternating current, or A.C.

25) Alternating current changes its direction back and forth in a circuit.

26) A transformer changes the voltage in an alternating current circuit.

Vocabulary Words

current	fuse
conductor	circuit breaker
insulator	ampere
direct current	volt
alternating current	ohm
closed circuit	resistance
open circuit	Ohm's Law
battery	galvanometer
dry battery	watt
wet battery	power
terminal	voltmeter
positive	ammeter
negative	switch
voltage	rheostat
parallel circuit	transformer
series circuit	primary winding
electromotive force	secondary winding

• CHAPTER REVIEW EXERCISES •

▶ Copy the sentences. Fill in the missing words.

1) An electric circuit with more than one path is called a ____ circuit.

2) Electric power is measured in ____.

3) A copper wire is an example of a good ____.

4) A battery has two ____, positive and negative.

5) Your house can be protected from a fire by a ____.

▶ Write the letter of the best answer.

6) You can measure electromotive force by using
   a) an ammeter.
   b) a voltmeter.
   c) a transformer.

7) Which of the following is not a common type of battery?
   a) A   b) AA   c) AAA

8) Electricity that flows only in one direction is called:
   a) alternating current.
   b) amperes.
   c) direct current.

9) A type of circuit with only one path is called a:
   a) series circuit.
   b) parallel circuit.
   c) open circuit.

10) Three $1\frac{1}{2}$ volt batteries connected in series will have a combined voltage of:
   a) 3 volts.   b) $4\frac{1}{2}$ volts.   c) $1\frac{1}{2}$ volts.

Computer Programs

Type these programs into your computer to solve for different electrical components — current, voltage, and watts.

▶ *Solving for Current:*

```
NEW
10  REM SOLVING FOR CURRENT
20  PRINT "ENTER THE VOLTAGE "
30  INPUT E
40  PRINT "ENTER THE RESISTANCE "
50  INPUT R
60  PRINT INT(E/R * 10+.5)/10;" AMPERES"
70  END
```

▶ *Solving for Voltage:*

```
NEW
10  REM SOLVING FOR VOLTAGE
20  PRINT "ENTER THE CURRENT "
30  INPUT C
40  PRINT "ENTER THE RESISTANCE "
50  INPUT R
60  PRINT C * R;" VOLTS"
70  END
```

▶ *Solving for Watts:*

```
NEW
10  REM SOLVING FOR WATTS
20  PRINT "ENTER THE VOLTAGE "
30  INPUT E
40  PRINT "ENTER THE CURRENT"
50  INPUT C
60  PRINT E * C;" WATTS"
70  END
```

Run each program. When you enter information, type the numerical portion only. Do not type units of measure, such as *amperes* or *ohms*. After you get an answer, attach the appropriate unit of measure to the answer.

Chapter **12**

Magnets and Electromagnetism

Chapter Goals:

1. To list three kinds of magnets.

2. To name the two poles of a magnet.

3. To state the rules of magnetism.

4. To describe magnetic fields and lines of force.

5. To describe how a compass works.

6. To state the relationship between electricity and magnetism.

7. To make an electromagnet.

8. To list three devices made from electromagnets.

12.1 NATURAL MAGNETS

There is a type of stone which can be found in the earth that has the unusual ability to attract pieces of iron. It is called *LODESTONE*. Its chemical name is *iron oxide*.

Lodestone can attract pieces of iron.

Lodestone is a natural *MAGNET*, an object which can attract certain other substances. Magnets do NOT attract all substances. Magnets attract some substances more than others.

12.2 KINDS OF MAGNETS

Natural lodestone, of course, can come in any size or shape, but most magnets which are made have a definite shape. One common shape for magnets is the familiar horseshoe magnet.

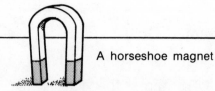

A horseshoe magnet

Other magnets are formed in the shape of bars, round cylinders, or even round doughnut shapes.

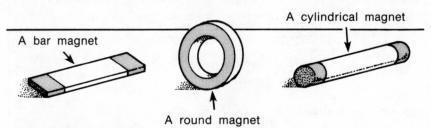

A cylindrical magnet

A bar magnet

A round magnet

12.3 POLES

Regardless of the shape, all magnets have a *NORTH POLE* and a *SOUTH POLE*. These are the points where most of the magnetism is concentrated.

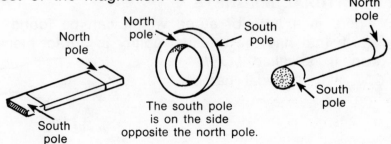

North pole

North pole

North pole

South pole

South pole

North pole

South pole

The south pole is on the side opposite the north pole.

It is not possible to tell whether the end of a magnet is a north pole or a south pole simply by looking at the magnet. The type of pole must be determined by testing. One way of telling is by placing the magnet close to another magnet whose poles are already known. By observing whether the magnets attract or *REPEL* (push away), you can determine the type of pole it is. Scientists use the following rules to determine poles.

Rules of Magnetism

1. Poles of the opposite type will attract each other.

2. Poles of the same type will repel each other.

In the following diagram, the magnets will attract each other because the poles are opposite.

North and south are opposites.

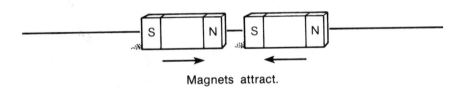

Magnets attract.

If one of the magnets is turned around, the magnets will repel each other.

Two south poles   Two north poles

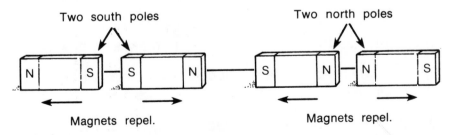

Magnets repel.   Magnets repel.

To determine the poles of an unknown magnet then, the scientist places one end of the unknown magnet against the pole of a magnet which is already labeled N and S.

In the diagram, if the unknown end of the magnet is repelled, it must be a south pole. If it is attracted, it must be a north pole.

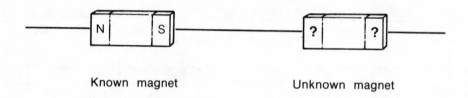

Known magnet   Unknown magnet

▶ *Practice.* Answer these questions.

1) What is lodestone?

2) What are three shapes of magnets?

3) How can you use magnets around your home? List two ways.

4) What are the two laws of magnetism?

5) What does *repel* mean?

6) What are the names of the two poles of a magnet?

7) How can you determine the poles of an unknown magnet?

8) What is the chemical which makes lodestone?

9) If two S poles are placed together, what will happen?

10) If two N poles are placed together, what will happen?

12.4 MAGNETIC FIELDS

Surrounding all magnets is a *MAGNETIC FIELD*, made of invisible lines of magnetism. This magnetic field is responsible for the attraction and repulsion of magnets.

Although you cannot see magnetic fields, you can see what they do, or their *effects*, by performing the following experiment.

Take a bar magnet and place it on a flat surface. Place a sheet of writing paper over the magnet.

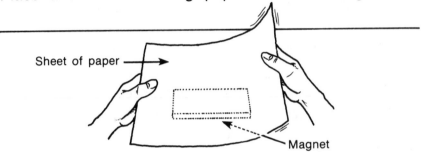

Sheet of paper ⟶

Magnet

Obtain some iron filings (small bits of iron) and sprinkle the iron filings on top of the paper. The iron filings will tend to line up in a definite pattern according to the presence of the magnetic lines of force.

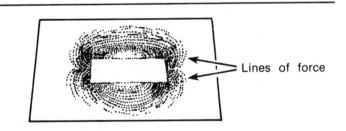

Lines of force

The magnetic lines of force appear to come out of one of the poles of the magnet and curve around to the other pole. All of the lines of force are called the magnetic field.

INVESTIGATION 10

Purpose: To observe the lines of force around various magnets.

A. Materials:

Quantity:	*Item Description:*
2	Horseshoe magnets
2	Bar magnets
2	Round magnets
1 box	Iron filings
3 sheets	Paper

B. Procedure

1) Place one magnet of each kind on a flat surface and cover with a sheet of paper.

2) Sprinkle some of the iron filings on each of the pieces of paper. DO NOT *POUR* the iron filings. It is best to sprinkle lightly from a height of about 1 foot.

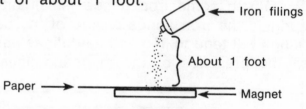

3) Observe the lines of force.

4) Record your observations in a data table like the one shown.

5) Place the bar magnets end to end with like poles together.

6) Again place a sheet of paper on the magnets and sprinkle with iron filings.

7) Observe the lines of force and record your observations in the data table.

8) Now reverse the poles so that opposite poles are end to end.

9) Sprinkle with iron filings.

10) Observe the lines of force. Record your observations on the data table.

11) Repeat these procedures with the horseshoe magnets.

12) Make a drawing for each of the following arrangements of magnets.

13) Write a description of the patterns produced by each arrangement.

C. Data Table

Magnets	Description of the Lines of Force
1) Horseshoe magnet	
2) Bar magnet	
3) Round magnet	
4) Horseshoe magnets with like poles end to end.	
5) Horseshoe magnets with unlike poles end to end.	
6) Bar magnets with like poles end to end.	
7) Bar magnets with unlike poles end to end.	

12.5 THE CAUSE OF MAGNETISM

If we could look at a piece of unmagnetized material under a powerful microscope, the atomic and molecular particles would not be in any regular pattern. They might look like the following diagram.

← Particles in random order

In a material that is magnetized, the particles align themselves into groups called *DOMAINS*, as in this diagram.

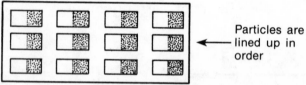

← Particles are lined up in order

If a magnet is broken into two parts, each of the two separate pieces will become magnets. Each of the two magnets will have north and south poles of its own.

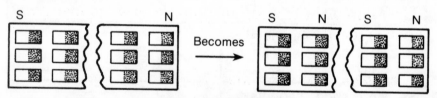

Becomes →

12.6 HOW TO MAKE A MAGNET

There are many ways to make a magnet. One of the simpler ways is to take an iron nail and stroke it with a magnet. Start by holding one of the poles of the magnet at the head of the nail. Slowly stroke the magnet down the length of the nail. After repeating this action four or five times, the nail will become a magnet.

12.7 HOW TO DESTROY MAGNETISM

Breaking a magnet into two parts does not destroy the magnetism. Instead, it produces two completely independent magnets. It is, however, possible to destroy the magnetism in a magnet. Anything that disturbs the magnetic domains will *DEMAGNETIZE*, or destroy, the magnetism of the magnet. Two common ways to demagnetize a magnet are by heating it, and by striking it with a hard blow.

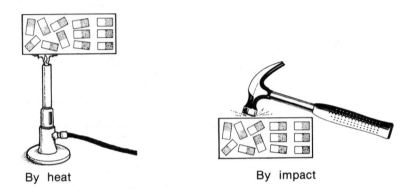

By heat   By impact

Destroying magnetism

If you want to keep magnetism in a magnet, avoid heating or dropping it!

▶ *Practice.*   Answer these questions.

1) What is a magnetic field?

2) How can you observe a magnet's lines of force?

3) How are the particles of a non-magnetized material arranged?

4) How are the particles of a magnetized material arranged?

5) What is the result of breaking a magnet into two pieces?

6) How can you destroy the magnetism in a magnet?

12.8 DOES MAGNETISM PENETRATE ALL MATERIALS?

There are many materials that are not magnetic. The following table lists some items that are magnetic and some that are not.

Substance	Magnetic?
Wood	No
Plastic	No
Iron	Yes
Rubber	No
Cobalt	Yes
Paper	No
Nickel	Yes
Copper	No
Leather	No
Steel	Yes
Gold	No
Silver	No
Glass	No

▶ *Practice*.

1) Divide the substances listed above into two groups — magnetic and non-magnetic. List each substance in the proper group.

2) Which group has more substances?

3) Is there any common property that all magnetic substances have?

INVESTIGATION 11

Purpose: To discover what materials magnetism will pass through and what materials will not allow it to pass through.

A. Materials

Quantity:	Item Description:
1	Paper clip
1	Piece of string
1	Ring stand
1	Ring clamp
1	Bar magnet
1	Paper
1	Piece of copper (any size)
1	Piece of plastic
1	Aluminum foil
1	Piece of glass
1	Piece of wood

B. Procedure

1) Set up the paper clip, ring stand, clamp, and bar magnet as in the diagram.

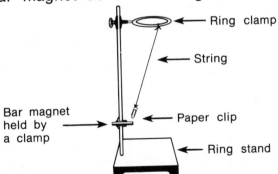

2) Test each object by placing it between the paper clip and the bar magnet. If the paper clip still is attracted to the magnet, then the material will allow magnetism to pass through it.

3) Record your results in a data table like the one shown below.

4) Try any other two materials which you may have, and record the results.

C. Data Table

Material	Passes Magnetism? (Yes or No)
Paper	
Copper	
Plastic	
Aluminum Foil	
Wood	
Glass	

12.9 THE EARTH AS A NATURAL MAGNET

The earth, as you have learned, contains natural magnets called lodestone. In addition to containing natural magnets, the earth actually is a giant magnet itself. It is because of this natural magnetism that the object known as a COMPASS works. A compass is a device containing a freely rotating needle which will point in a north-south direction. It is used to find directions on the surface of the earth.

A compass is used to find direction.

The compass needle has a north pole and a south pole located at either end of the needle. The earth has a magnetic field and north and south poles which will attract the compass needle.

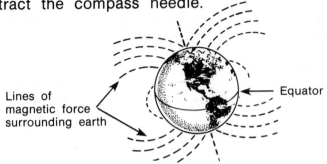

The magnetic poles of the earth are not located at the true geographic north and south poles. A compass, therefore, does not point to the real north pole, but to the magnetic north pole. When using a compass for accurate work, allowance must be made for the difference in location of true north and magnetic north.

12.10 MAGNETISM AND ELECTRICITY

Although magnetism and electricity are not the same type of energy, they are related. Where there is electricity flowing, you will usually find a magnetic field. You can prove this for yourself by using a compass as an electricity indicator. Place the compass near a wire which is carrying electricity. The compass needle will change its direction as it comes closer to the wire.

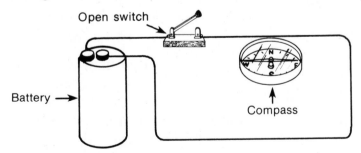

The compass needle will move when current is turned on or off by the switch.

A changing electric current produces a magnetic field. Electricity, in fact, can be used to make a type of magnet called an *ELECTROMAGNET*. An electromagnet acts just as if it were a normal magnet except that it is only magnetic as long as there is an electric current flowing. Here is a diagram of how to construct an electromagnet.

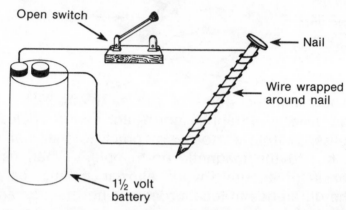

The electromagnet should be made with a large nail, some common bell wire, and a 1½ volt dry cell battery. The switch, while not necessary, is good to have because the electromagnet will heat up quickly. The circuit can be shut off, then, by opening the switch. If you leave the switch closed, the battery will quickly wear down due to excessive current drain.

The strength of the electromagnet depends on a number of factors. The higher the voltage of the battery, the more powerful is the electromagnet. Also, the strength of the electromagnet will be increased by placing more turns of wire on the nail. You can experiment with electromagnets on your own. Try different voltages of batteries. Increase or decrease the number of turns of wire. Try different objects to wrap the wire around. You can measure the strength of your electromagnets by counting how many paper clips they will pick up.

12.11 DEVICES USING ELECTROMAGNETS

Many modern day appliances use electromagnets. Microphones, speakers, earphones, and telephones all contain electromagnets. Their purpose is to change electrical currents into sound waves.

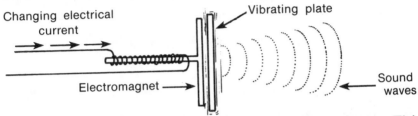

Changing electrical current

Vibrating plate

Electromagnet

Sound waves

The electrical current changes in the wires. This causes the electromagnet to become strong, then weak, then strong, then weak, and so forth. As the electro-magnet becomes stronger and weaker, the plate is moved in step with the electricity. When the plate moves or vibrates, it creates sound!

The telegraph also uses an electromagnet. The telegraph was invented in the 1830's to send messages over long distances. A telegraph operates by opening and closing an electric circuit. As the circuit is closed by a key, the electric current causes an electromagnet to be energized. The electromagnet, in turn, pulls on a metal bar called a sounder. The sounder makes the familiar dots and dashes of the Morse code used in telegraphy.

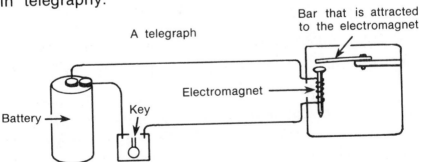

A telegraph

Bar that is attracted to the electromagnet

Electromagnet

Battery

Key

To have two-way communication, it is necessary to have two telegraphs.

SUMMARY OF CHAPTER 12

1) Natural magnets are called lodestone.

2) A magnet is an object which will attract certain other objects.

3) Magnets come in many shapes and sizes.

4) Magnets have two poles — a north pole and a south pole.

5) Like magnetic poles will repel, or push away.

6) Unlike magnetic poles will attract.

7) Magnets have a magnetic field around them made up of lines of force.

8) You can see the lines of force of a magnet by covering the magnet with a piece of paper and sprinkling iron filings on it.

9) The particles of magnetic materials align themselves into domains.

10) Magnetism can be destroyed by heat and impact.

11) The earth is a natural magnet.

12) The magnetic poles are not located at the geographic poles of the earth.

13) A compass will point in the direction of the magnetic poles.

14) Magnetism and electricity are closely related.

15) Electricity can be used to produce an electromagnet.

16) An electromagnet can be strengthened by increasing the battery voltage or by increasing the number of turns of wire.

17) Many modern day appliances use electromagnets.

• CHAPTER REVIEW EXERCISES •

Vocabulary Words

lodestone
magnet
horseshoe magnet
bar magnet
north pole
south pole
attract

repel
magnetic field
domain
demagnetize
compass
earth's magnetic poles
electromagnet

▶ Copy the sentences. Fill in the missing words.

1) A natural magnet found in the earth is ____.

2) Two magnetic north poles will ____ each other.

3) A ____ can be used to find direction.

4) The particles of a magnet are aligned in ____.

5) Opposite magnetic poles ____ each other.

▶ Write the letter of the correct answer.

6) An object can be demagnetized by
   a) placing it in water. b) cooling. c) heating.

7) The earth's magnetic north pole
   a) is located at the geographic north pole.
   b) is not located at the geographic north pole.
   c) will not affect a compass.

8) A north pole of a magnet is brought close to the pole of another magnet, which is then attracted.
   a) The unknown pole is a south pole.
   b) The unknown pole is a north pole.
   c) The unknown pole cannot be determined.

9) Which of the following is magnetic?
   a) paper   b) copper   c) steel

10) The "lines" which are visible around a magnet when sprinkled with iron filings show the
   a) lines of force. b) domains. c) lodestone.

Computer Program

▶ The following program will print a vocabulary quiz. The word will be displayed for a short period. Then you will be asked to spell it correctly. Up to ten words can be included in the DATA statements. If you do not have enough time to study the words, you can change the value in line 320 from 4500 to 5000 or 6000. Experiment for best results.

```
10 CLS
20 N=5:C=0
30 DATA MAGNET,ATTRACT,REPEL,FIELD,DOMAIN
40 FOR I=1 TO N
50 READ A$(I)
60 NEXT I
70 CLS
80 IF C=N THEN 290
90 PRINT "STUDY THIS SPELLING WORD."
100 PRINT "YOU WILL HAVE ABOUT 10 SECONDS."
110 C=C+1
120 PRINT
130 PRINT A$(C)
140 GOSUB 320
150 CLS
160 PRINT "TYPE THE CORRECT SPELLING."
170 PRINT
180 INPUT B$(C)
190 IF B$(C)=A$(C) THEN 250
200 PRINT
210 PRINT B$(C);" IS MISSPELLED."
220 PRINT A$(C);" IS THE CORRECT SPELLING."
230 GOSUB 310
240 GOTO 70
250 PRINT B$(C);" IS SPELLED CORRECTLY."
260 GOSUB 310
270 IF C< >N THEN 70
280 PRINT
290 PRINT "END OF TEST"
300 END
310 FOR T=1 TO 4500:NEXT T:RETURN
320 FOR T=1 TO 4500:NEXT T:RETURN
```

Appendix — Table of the Elements

Element Name	Symbol	Atomic Number
actinium	Ac	89
aluminum	Al	13
antimony	Sb	51
argon	Ar	18
arsenic	As	33
astatine	At	85
barium	Ba	56
beryllium	Be	4
bismuth	Bi	83
boron	B	5
bromine	Br	35
cadmium	Cd	48
calcium	Ca	20
carbon	C	6
cerium	Ce	58
cesium	Cs	55
chlorine	Cl	17
chromium	Cr	24
cobalt	Co	27
copper	Cu	29
dysprosium	Dy	66
erbium	Er	68
europium	Eu	63
fluorine	F	9
francium	Fr	87
gadolinium	Gd	64
gallium	Ga	31
germanium	Ge	32
gold	Au	79
hafnium	Hf	72
helium	He	2
holmium	Ho	67
hydrogen	H	1
indium	In	49
iodine	I	53
iridium	Ir	77
iron	Fe	26
krypton	Kr	36
lanthanum	La	57
lead	Pb	82
lithium	Li	3
lutetium	Lu	71
magnesium	Mg	12
manganese	Mn	25

Element Name	Symbol	Atomic Number
mercury	Hg	80
molybdenum	Mo	42
neodymium	Nd	60
neon	Ne	10
nickel	Ni	28
niobium	Nb	41
nitrogen	N	7
osmium	Os	76
oxygen	O	8
palladium	Pd	46
phosphorus	P	15
platinum	Pt	78
polonium	Po	84
potassium	K	19
praseodymium	Pr	59
promethium	Pm	61
protactinium	Pa	91
radium	Ra	88
radon	Rn	86
rhenium	Re	75
rhodium	Rh	45
rubidium	Rb	37
ruthenium	Ru	44
samarium	Sm	62
scandium	Sc	21
selenium	Se	34
silicon	Si	14
silver	Ag	47
sodium	Na	11
strontium	Sr	38
sulfur	S	16
tantalum	Ta	73
technetium	Tc	43
tellurium	Te	52
terbium	Tb	65
thallium	Tl	81
thorium	Th	90
thulium	Tm	69
tin	Sn	50
titanium	Ti	22
tungsten	W	74
uranium	U	92
vanadium	V	23
xenon	Xe	54
ytterbium	Yb	70
yttrium	Y	39
zinc	Zn	30
zirconium	Zr	40

Glossary

acceleration: The change in speed divided by the change in time.

acid: A special compound containing hydrogen.

acoustics: The science of sound.

alchemist: A person who tried to change various substances into gold and other precious metals.

alloy: A mixture of two or more metals.

alternating current: A type of electricity that continuously changes polarity.

ammeter: A device that is used to measure the current in an electrical circuit.

ampere: The unit of electrical current.

analysis reaction: A reaction in which a compound is broken down into the elements of which it is made.

atom: The smallest particle of an element.

atomic mass number: A number that is equal to the number of protons plus the number of neutrons in an atom.

atomic number: A number that is equal to the number of protons in an atom.

attract: To pull together.

average speed: The total distance traveled divided by the total time taken to travel that distance.

balance scale: An instrument used to measure mass or weight.

base: A special compound containing the OH radical.

battery: A device that changes chemical energy into electrical energy.

beaker: A glass or plastic cylinder used to hold liquids.

boiling point: The temperature at which a liquid begins to change into a gas.

calorie: The amount of heat needed to raise the temperature of one gram of water by one degree Celsius.

Celsius: The temperature scale used in scientific work.

centimeter: A unit of length in the metric system. Equal to 1/100 of a meter.

chemical equation: A statement that uses symbols and formulas to describe a chemical reaction.

chemical formula: A group of chemical symbols telling what atoms are present in a compound and how many atoms of each element are in the compound.

chemist: One who studies chemistry.

chemistry: The study of matter.

circuit breaker: An electrical device that is designed to open a circuit and prevent a fire.

coefficient: A number placed before a chemical formula which indicates the number of molecules in a chemical equation.

compass: A device with a magnetic needle that always points north; used to find direction.

compound: A substance made of two or more elements combined chemically.

concave lens: A lens which is thinner in the middle than at the edges.

concave mirror: A mirror which curves in at the middle.

conduction: A method of heat movement caused by molecules bumping into each other.

conductor: An object that will conduct heat, sound, or electricity.

convection: A type of heat movement in liquids and gases.

convex lens: A lens which is thicker in the middle than at the edges.

convex mirror: A mirror that curves outward at the middle.

cubic centimeter: A cube one centimeter on each side. Equal to 1 milliliter.

current: The movement of electrons from one place to another.

cycle: One complete back-and-forth vibration.

decelerate: To go slower. A negative acceleration.

decibel: The unit used to measure the loudness of a sound.

demagnetize: To remove magnetism.

density: The mass of an object divided by its volume.

direct current: A type of electricity that flows only in one direction.

dissolve: To melt; to mix with a liquid; to pass into a solution.

double replacement reaction: A chemical reaction in which the atoms of two compounds exchange places with each other.

echo: A reflected sound wave.

effort force: The force that is applied to a simple machine.

electricity: One of the forms of energy.

electromagnet: A device that becomes a magnet when electricity is applied to it.

electromotive force: The force that pushes electrons through a circuit. It is measured in volts.

electron: A particle that revolves around the nucleus of an atom. It has a negative charge.

electron shell: Areas surrounding the nucleus of the atom which contain the electrons.

element: One of 92 natural substances which are the basic building blocks of matter.

energy: The ability to do work.

evaporate: To change from a liquid to a gas.

Fahrenheit: The temperature scale commonly used in the United States.

family of elements: A group of elements that have similar properties.

farsighted: Unable to see clearly, objects that are close up.

fixed pulley: A pulley that is attached to something and therefore does not move.

focus: The point at which the rays of light come together to form an image.

foot-pound: The unit of work in the English system of measurement.

force: A push or a pull.

formula: An abbreviation for compounds telling what atoms and how many atoms of each element the compound contains.

freezing point: The temperature at which a liquid changes to a solid.

frequency: The number of times a wave vibrates in one second.

friction: A force that opposes motion.

fulcrum: The point around which a lever rotates.

fuse: An electrical device that is designed to melt, thereby opening a circuit and preventing a fire.

galvanometer: A type of meter that measures very small voltages.

gas: One of the states of matter. Gases have no definite shape and no definite volume.

graduated cylinder: A round glass or plastic cylinder used to measure liquids.

gram: The unit of mass in the metric system.

gravity: The force of attraction between any two objects that have mass.

heat: A type of energy caused by the motion of molecules. Heat depends on the temperature and the amount of mass of an object.

horseshoe magnet: A type of magnet shaped like a horseshoe.

image: A likeness of something that can be seen in a mirror or focused by a lens.

inclined plane: A type of simple machine that is shaped like a ramp.

index of refraction: A number that measures how much an object will refract a light wave.

inert: Will not combine with other substances under ordinary conditions.

insulator: An object that will not allow heat, sound, or electricity to travel through it.

ion: An atom that has a charge due to gaining or losing an electron.

isotope: A different form of an element. An isotope has the same number of protons and electrons as the ordinary element. However, it has a different number of neutrons.

joule: The metric unit of work.

kinetic energy: The energy of motion.

laboratory: A place where scientists conduct experiments.

lever: A type of simple machine consisting of a bar which is free to rotate around a central point.

light: A form of energy that can be seen by our eyes.

linear: Pertaining to length.

liquid: One of the states of matter. A liquid has a definite volume but no definite shape.

liter: The unit of liquid volume in the metric system.

lodestone: A naturally magnetic stone.

magnet: An object that will attract certain types of metals.

magnetic field: Lines of invisible magnetism that surround a magnet.

magnify: To enlarge an image.

mass: A measure of how much matter an object contains.

matter: Anything that has mass and takes up space.

mechanical advantage: The number of times a machine multiplies the effort force.

melting point: The temperature at which a solid changes into a liquid.

meniscus: The rounded top of a liquid seen in a graduated cylinder when measuring volume.

metal: A group of elements placed on the left side of the periodic table. Metals are usually shiny and are good conductors of heat and electricity.

meter: The standard unit of length in the metric system.

metric system; A system of measurement whose units are based on the number 10.

microscope: An optical device used to magnify small objects.

millimeter: A unit of length in the metric system. Equal to 1/1,000 of a meter.

mirror: An object that will reflect light.

molecule: The smallest particle of a compound. Made of one or more atoms.

movable pulley: A pulley that is not attached to a stationary object and is therefore free to move.

nearsighted: Unable to see clearly, objects that are far away.

neutron: A particle found in the nucleus of the atom. It has no charge.

newton: The unit of force in the metric system.

noble gases: A group of gases that do not react with other substances under ordinary conditions.

non-metal: Elements found on the right side of the periodic table. They are not good conductors of heat and electricity.

north pole (of a magnet): One of the two ends of a magnet.

nucleus: The central part of the atom which contains the protons and neutrons.

ohm: The unit of electrical resistance.

Ohm's Law: An equation which relates the current, voltage, and resistance in a circuit. The current is equal to the voltage divided by the resistance.

ohmmeter: A device that is used to measure the resistance in an electrical circuit.

optics: The science of light.

parallel circuit: A type of circuit having more than one path for the current.

periodic table: A table that lists the elements in order of increasing atomic number.

photons: The name given to particles of light.

pitch: The highness or lowness of a sound.

plane mirror: A flat mirror.

potential energy: The energy of position. Stored energy.

pound: The unit of force and the unit of weight in the English system of measurement.

power: A measure of the amount of electricity used.

precipitate: A solid that settles to the bottom of the container during a chemical reaction.

prism: A triangular piece of glass or plastic that will break light up into the colors of the spectrum.

product: A substance that is formed during a reaction. A product is placed on the right side of the arrow in a chemical equation.

property: A quality or characteristic that helps to identify a substance.

proton: A particle found in the nucleus of an atom. It has a positive charge.

pulley: A simple machine consisting of a wheel with a rope, string, or chain that wraps around the wheel.

radiation: One method by which heat travels. Heat travels from the sun to the earth by radiation.

radical: A group of two or more atoms that act like one element.

reactant: A substance that enters into a reaction. Found on the left side of the arrow in a chemical equation.

reaction: The change of one substance into a different substance.

real image: An image that can be projected.

reflect: To bounce back sound or light waves.

refract: To bend.

repel: To push apart.

resistance: The opposition to the flow of electricity in a circuit. The resistance is measured in ohms.

resistance force: The object that is to be moved by a machine.

resistor: Any device that offers a resistance to the flow of electricity.

rheostat: An electrical device that can vary the amount of resistance in a circuit. It can control the brightness of a bulb.

scale: An instrument used to measure mass or weight.

series circuit: A type of circuit having only one path for the current.

simple machine: A device that increases force, increases the speed of an object, or changes the direction of a force.

simple replacement reaction: A chemical reaction in which an atom changes places with another atom in a compound.

solid: One of the states of matter. A solid has a definite shape and a definite volume.

solute: The substance which dissolves in a solution.

solution: A mixture formed when one substance is dissolved in a liquid.

solvent: The liquid in which a substance dissolves.

sonar: A device that measures underwater distances by using reflected sound waves.

sound: A form of energy that can be heard by our ears.

south pole (of a magnet): One of the two ends of a magnet.

spectrum: The colors of the rainbow which make up white light.

speed: The distance traveled per unit of time.

spring scale: A device used to measure forces.

static electricity: A type of electricity usually caused by friction.

subscript: Used in formulas to tell the number of atoms of an element that are found in a compound. They are small numbers written to the right of the element to which they apply.

switch: An electrical device placed in a circuit to turn the electricity on or off.

symbol: An abbreviation for the name of an element.

synthesis: A chemical reaction in which two or more atoms combine to form one compound.

telescope: An optical device used to study heavenly bodies.

terminal: A part of a battery that is connected to a circuit. Batteries have two terminals, the positive and negative.

temperature: A measure of the average motion of the molecules in a substance. Temperature is measured with a thermometer.

thermometer: A device that measures temperature.

transformer: An electrical device that can change the voltage in an alternating current circuit.

vacuum: Refers to a space that contains no matter.

vibrate: To move back and forth.

virtual image: An image that cannot be projected.

volt: The unit of electromotive force.

voltage: Electrical pressure that is measured in volts.

voltmeter: A device that measures voltage in a circuit.

watt: The unit of electrical power.

weight: A measure of how strongly the earth's gravity attracts something.

work: A force moving through a distance. Found by multiplying the force by the distance moved.

Index